After Duwagan

Deforestation, Succession, and Adaptation in Upland Luzon, Philippines

J. Peter Brosius

Michigan Studies of South and Southeast Asia Number 2
Center for South and Southeast Asian Studies
The University of Michigan

Library of Congress Catalog Card Number: 89-81764

ISBN 0-89148-061-7

Printed in the United States of America

To the memory
of my grandfather

Radford H. Lippy

CONTENTS

TABLES

FIGURES

MAPS

FOREWORD

Karl L. Hutterer

It is rare for our earliest research efforts to be published in monographic form. As a rule, we are ourselves aware of their shortcomings, and it is the sign of a good scholar to perceive such imperfections all the more keenly. However, the current study, first produced as a master's thesis in anthropology, stands out among its peers, presenting information and analyses that strongly warrant its publication. I am pleased that Peter Brosius yielded to our urgings and agreed not only to present to the public a work that he feels is not fully mature but that he was willing to invest considerable energy in revising and preparing the manuscript. In my view, this book makes important contributions to our understanding of the "Negrito" peoples in Southeast Asia, to the anthropological and ecological study of systems of shifting cultivation and, in general, to the interpretation of landscapes that have been profoundly affected by human activity.

Since the beginning of this century, a number of ethnographies have been published about Philippine peoples generically grouped under the label "Negrito." While all these studies have reported mixed subsistence strategies, most have focused on the hunting and gathering component. The tendency has been to assume that hunting and collecting in the tropical forest once characterized the subsistence economy of all Negrito groups as an exclusive strategy, and that wage labor, trade of forest products and agriculturally produced food staples, and swiddening represent relatively recent acculturation to external pressures.

In light of these assumptions, Brosius' report on the Ayta comes as something of a surprise. The Ayta are dedicated swiddeners (although, like most others, they complement farming with foraging, particularly for protein foods). They have been farmers for such a long time that the memory of an economy dominated by hunting and gathering has retreated into the mythological past, and their involvement with farming has been so extensive that they have become major contributors to the degradation of their own environment. Little is known about the long-term past of the Ayta as a distinctive ethnic group and it will probably be difficult to go much further in reconstructing it than Brosius has. Given this lack of information, it is difficult to draw well-founded conclusions about the implications of these ethnographic facts for the social evolution and social history of the northern Philippines, or even for Negrito populations specifically. Nonetheless, the ethnographic data presented here should stand as a warning against simplistic assumptions regarding the nature and role of Negrito peoples in the human history of the Philippines and Southeast Asia.

It is interesting to note that, while the Ayta have been dedicated swiddeners for as long as they can remember, probably for hundreds of years, they retain a strong interest in a bow-and-arrow technology. The fact that they generally travel with bow-and-arrow, and the enormous diversity of projectile point forms, may well indicate a cultural retention of elements that once had important functions in their subsistence technology but serve today quite different purposes. These observations are witness to the fact that material relations are by no means the only factor determining a people's cultural life but that this is centered around sets of meanings having an impetus of their own.

Although it was not intended that way, Brosius' work is also an instructive case study of the environmental ethic of a tribal group. As A.T. Rambo cogently observed in a study of Malaysian Semang,* "primitive" groups appear to be conservative in their approach to the environment mainly by virtue of their low population densities, small population size, and low energy technologies, resulting in most cases in a low level of impact on their habitats. Daily decisions governing interactions with the environment are dictated more by ad hoc concerns about meeting requirements for food, shelter, and minimal comfort than by long-range considerations of environmental conservation

---

* A.T. Rambo, *Primitive Polluters: Semang Impact on the Malaysian Tropical Rain Forest Ecosystem*, Anthropological Papers No. 76 (Ann Arbor: Museum of Anthropology, University of Michigan, 1985).

and resource sustainability. Considered close-up and from the perspective of conservation ecology, the approach to manipulation of natural resources of "primitive" groups may in some cases be rather profligate. Under these conditions, even relatively minor changes in sociopolitical conditions or in technology--such as the introduction of guns--can have far-reaching effects, setting off a spiral of environmental degradation from which it is difficult to recover.

One of the main foci of Brosius' work is, of course, the ecology of swidden farming in an environment that has been degraded from a forest climax to one covered by grassland and other depauperate secondary associations. In the original thesis, he was concerned with two opposed notions of swiddening: a romantic concept in which swidden farming is seen to replicate the ecological structure of tropical forests in an agricultural context, and a bureaucratic (and political) concept that views it as the chief cause of the destruction of tropical forests. While the published version avoids the polemics of the original thesis, it still serves to point out the diversity of swidden systems and the diversity of conditions under which swidden farming may be practiced. Both aspects must be taken into account when making judgments about the benefits of, or damage incurred by a particular swidden economy.

Brosius himself investigates a case of what might superficially be referred to as grassland swiddening. The shifting cultivation of grasslands has seldom been studied, even though it is common in some parts of the world, including areas of Southeast Asia. Calling attention to this aspect of swidden farming is a major service. In his analysis, Brosius employs the construct of ecological succession theory, an approach that has seen surprisingly little use in ecological studies in anthropology. Again, new ground is broken in this monograph.

In the ecological discussions presented below, among all the natural science information about species, climates, landscape, and so on, some readers will wish for a fuller picture of the Ayta as individuals and as a people. Indeed, some might wonder why this book is being published in a series on Southeast Asian studies, a field traditionally concerned with culture, religion, history, and the fine arts. I would turn the question around: why has Southeast Asian studies paid so little attention to the environment in which people live and with which they interact? It is a trite but true observation that cultural and social structures, as well as cultural meanings, cannot be fully understood unless they are seen in the context of these interactions--although it is also true that an ecological perspective cannot by itself account for cultural forms and contents. The simple fact is that the relations between the material basis of human existence and its cultural construction are not well understood. It is precisely for that reason that they are worth studying. The present work might be considered one-sided in its attention to ecological relations in a material context but within that framework it is well done. Brosius is successful in using ecological succession theory in explaining the creation of an anthropogenic environment and thus adds an important new tool to our attempts at understanding the dynamic interactions between human communities and the world they live in. In this he makes a contribution that is relevant beyond this specific study and even beyond the geographical boundaries of the Southeast Asian region.

PREFACE

This monograph is based on research carried out from January to August 1980 among the Ayta Negritos of southern Zambales Province, Philippines. The present work was originally submitted as a masters thesis at the University of Hawaii, Department of Anthropology, and was completed in 1981. Though I have been encouraged to publish this for some time, only recently have I had the opportunity to return to the original manuscript in order to make the revisions that I felt were necessary. The present text differs in many particulars from the original, but none of the revisions are in any way substantive. I have eliminated substantial portions of text, those which now seem to have been tangential to the central argument or which seem to have been rather pedantic in intent. A large number of citations have been omitted, paragraphs have been rearranged, and repetitive sections eliminated. Some points in the original text were not very clear, and I have tried to clarify these where necessary. Further, I have made numerous stylistic corrections in the original prose. The first chapter has undergone the most extensive revisions. The tone of Chapter 1 in the original document was at times programmatic and overly dismissive of earlier work: I have revised this considerably. I have also merged what was originally a separate chapter on disturbance into this first chapter. Chapter 2 (Ayta Subsistence, Settlement and Social Organization) has also been moved from its position (as Chapter 5) in the original text. Other than these changes, very little new text has been added. I would hope that the result of these revisions is a more focused and readable text which will more clearly convey the points I wish to make.

It is never easy going back to something written many years before. Since writing the original manuscript my interests have shifted both geographically and theoretically. I have most recently worked in Malaysia, having spent three years working with Penan hunter-gatherers in Sarawak on the island of Borneo. Though I retain a strong interest in ecological issues, my present work is rather far removed from such concerns. More importantly, I have become thoroughly unconvinced of the utility of a purely behavioristic approach to ethnographic observation, an approach which is very strongly evident in this work. Since writing this I have become much more interested in issues of culture and in what might generally be termed the constitution of society. Readers will notice that such concerns are completely absent in the following. In focusing so purely on the physical environment and on the details of human activities affecting this upland ecosystem, the Ayta themselves are nearly invisible. This has been the most frequent criticism of this work, one with which I am presently in complete agreement. When I wrote this, I was largely opposed to any notion of culture, emphasizing instead the primacy of behavior. The absurdity of this view is clear to me now, and I recognize that it caused me to ignore a great deal of ethnographic significance. In addition to shortcomings of an ethnographic nature, I also recognize some that are more purely empirical. That is to say, there are several types of data I did not collect. I did not, for instance, collect soil samples, nor were plants identified from voucher specimens (for which I lacked the necessary budget). Other shortcomings are merely the result of a limited period in the field--eight months--and the fact that I was rather mobile. Of course, for data on things such as temperature and rainfall, short-term measurements would have been meaningless.

Though I clearly recognize the shortcomings of this work, I feel that it makes a contribution to the literature on the anthropogenic disturbance of landscapes. I would judge its primary value to be the focus on succession theory as a way of interpreting the genesis of anthropogenic landscapes, incorporating the historical dimensions of change in a theoretically parsimonious manner. While this would seem particularly useful as a way of looking at swidden systems, it is applicable to any situation wherein landscapes are being transformed by human activity. Such a model provides both an agenda for research and a way to interpret results. By understanding (a) the context in which disturbance occurs, (b) the types of disturbances occurring, and (c) the parameters of disturbance, we can begin to understand the dynamics of ecosystems in which humans are primary actors.

Such a program has relevance to a number of present concerns. One of the most pressing is the need to understand the impact of human activities on tropical ecosystems and to be able to judge the sustainability of those ecosystems under different regimens of disturbance. Tropical forests in Southeast Asia and elsewhere are being destroyed at an alarming rate, and already huge areas have been deforested. Just since 1980 some of the last extensive stands of forest in the Philippines have fallen, and the same is occurring throughout Borneo, the Malay Peninsula, Sumatra and elsewhere: much of this is due to the activities of logging companies. At the time this manuscript was written, most Southeast Asian governments were unconditionally condemning shifting cultivation as a wasteful and destructive practice and the primary cause of deforestation. Many still do. Though in many cases the motivation for this is purely political, it is just as often due to a lack of understanding of such systems. Even those with a first-hand knowledge of swidden systems frequently do not recognize that there is a range of factors impinging on the sustainability of such systems. I recently attended a conference concerned with the issue of biodiversity in Borneo. On a number of occasions participants argued as to whether Bornean swidden systems were responsible for deforestation. In Sarawak, the government and conservationists are taking sides over this issue as well. The problem in such cases is a lack of understanding of the factors that determine the type of impact a particular system will have: factors as simple as whether swiddeners clear fields from primary forest (as do Iban) or from fallow growth (as do Kayan and Kenyah).

There is clearly a need for clarification of the factors which determine the outcome of environmental disturbance. Recently there has been considerable work done by local researchers throughout Southeast Asia and, in part because of this, a gradual reassessment of swiddening in some of these countries by government forest authorities, especially in the Philippines. Part of the purpose of this monograph was to provide a method whereby such systems could be assessed on a more empirical basis, rather than being condemned outright: simply stated, each case must be assessed individually. In the particular case with which this monograph is concerned, I found the Ayta swidden system to have been quite destructive, albeit in orchestration with other types of disturbance. But in showing this I also attempt to provide a way of interpreting and demonstrating empirically why this is so. This enables us to view swidden systems in a comparative context, allowing us to understand (a) not only how different swidden systems compare but how they are integrated with other types of disturbance in a particular setting, and (b) how, in similar settings, different types of disturbance compare (for instance, logging and swiddening). This may allow us to demonstrate why particular swidden systems are sustainable or, if they are not, why this is so. An understanding of factors of context and disturbance can direct us to those points which may cause particular systems to degrade an ecosystem, and thereby enable us to develop solutions. This applies not just to swidden systems but to all types of disturbance. Often, more than one type of disturbance is occurring in a particular ecosystem. I hope to have provided a theoretically coherent way of looking both at the range of disturbances occurring in a single ecosystem and of similar types of disturbances occurring in different ecosystems.

My research was stimulated by two extended visits to the Philippines in 1973 and 1974, during which several visits were made to Ayta settlements throughout Zambales and Bataan. The degree to which environmental change had apparently occurred in these areas made a deep impression on me. Prior to 1980 no ethnographic research had yet been done among the Ayta population inhabiting the upper western watershed of the Cabusilan Mountains (see Map 1), though a number of observers have passed through the area. Reed (1904), Schebesta (1952-57) and Fox (1953) were all concerned with groups in lower elevation areas to the west and north. Reed visited Aglao and Kabayan in 1903, and Smith, with a detachment of U.S. Marines, visited Lomboy in 1907 (Smith 1909). In 1938-39 Schebesta visited Kabayan, to the immediate west of the Lomboy Plain, and made a trip into Maagə'agə'. Garvan (1963) and Parker (1964) both visited groups on the eastern margin of the research area in Pampanga Province. Garvan may have travelled through the area, but his lack of locational identifications makes this impossible to determine. From October 1977 to April 1979 Hiromu Shimizu and Michiko Yamashita carried out ethnographic and linguistic research among Ayta in the village of Kakilingan, on the immediate southwestern edge of the research area (see Shimizu 1981, 1983; Shimizu and Yamashita 1982). Many individuals in Kakilingan are the immediate kin of those within the research area, and virtually all of Shimizu's and Yamashita's statements about Kakilingan Ayta apply to those among whom I worked in the higher elevations to the east.

The Ayta population within the immediate research area presently numbers about three hundred. Their language, a subdialect of Botolan Sambal, differs slightly from surrounding Negrito groups to the south, north and west. They have close affinities with Ayta on the eastern Pampanga watershed of the Zambales Mountains. All interviews were carried out in the Ayta dialect. Unless otherwise specified, further references to Ayta will refer only to those communities within the bounds of the research area.

My research strategy was to spend a month or more in a series of settlements or camps, in order to have the widest perspective on intragroup variability. Five areas were selected for primary focus: Maagə'agə', Bayabas, Ungot, Kalantas and Lomboy (see Map 2).

Outside sources were consulted as well. Primary among these were historical documents gathered from the Filipiniana Division of the National Library of the Philippines, the Philippine National Archives, the Ateneo de Manila Library and the American Historical Collection of the U.S. Embassy in Manila. Aerial photographs, vegetation surveys and logging maps from the Bureau of Forest Development were also consulted.

The Ayta do not name rivers or other features as such. Rather, names are given to specific sites, areas or districts. In this monograph rivers and other features may be given names not otherwise used by the Ayta, for purposes of clarity and specificity. This will be indicated where done. Frequently here I refer to the "research area," while at other times I speak of the "Cabusilan area," "upland Zambales," "southern Zambales," "western Luzon," and so forth. When I speak of the research area, this is in reference to the roughly triangular area in which my work was concentrated, and is used where more precision is desired, such as when I refer to percentages of area under forest cover, under grassland, or burned annually. When it is sufficient to speak in more general terms I refer to the "Cabusilan area" or employ another geographical referent. The word *Cabusilan* is the name given on maps for this particular part of the Zambales mountain chain, but is not used by Ayta. I employ it as a matter of convenience.

Finally, one brief point of clarification. The Ayta among whom this work was carried out are a Negrito people. Be this as it may, I avoid the word *Negrito* whenever possible. I am uncomfortable with this word primarily because it brings race to the forefront when, in the present case, it is absolutely irrelevant except perhaps in the speculative arena of history. Also the word *Negrito* is a diminutive form. While the diminutive may simply refer to stature, it would also seem to carry vaguely racist connotations. Though I try to avoid it wherever possible, this is difficult because the word is so entrenched in the literature. I employ the word *Ayta* as much as possible, both because it is free of any connotations and because it is what they call themselves.

The title *After Duwagan* is derived from a myth told by Pan-Pungpung about a man, Duwagan, who was one of the first ancestors, the *onan tao'*. Duwagan was an *amukok*, a person who had no swidden, who hunted and wore bark cloth, what we might term a hunter-gatherer. The following is an account of what has happened since Duwagan's time.

J.P.B.
1989

ACKNOWLEDGMENTS

It is impossible for me to adequately thank all those whose help and support have made it possible to bring this work to some semblance of completion. Special thanks must go to the Charles A. Lindbergh Fund, Inc., whose financial support made possible the research upon which this work is based. Their support is gratefully acknowledged. The National Museum of the Philippines, with which I was affiliated in the Philippines, are also to be thanked; Alfredo Evangelista and Jesus Peralta were particularly helpful. Also in the Philippines, I wish to thank Dr. Romeo Bruce (University of the Philippines), Antonio Rodriguez, the Benquet Consolidated Mining Corporation, and the Bureau of Forest Development. All provided valuable assistance and information. The librarians of the Filipiniana Division of the National Library of the Philippines were also extremely helpful. For "logistical" support in the Philippines I wish to thank Rufino and Aloma Tima, Hank and Theresa Schilling, and Rebecca Estioko. Their hospitality is gratefully acknowledged.

My committee at the University of Hawaii--Richard Gould, Bion Griffin, Alice Dewey and Richard Lieban--were very supportive throughout the period of research and writing. Many of the ideas presented here are the result of long and frequent discussions with Dr. Gould. Bion Griffin, through many seminars and discussions, has greatly influenced my perception of the literature on hunter-gatherers, Philippine ethnography in general, and Negritos in particular. He has also been justly critical of the manner in which I sometimes present my ideas and has been a great influence in this respect.

Many other people at the University of Hawaii contributed greatly to this work. Most of the ideas presented here were first discussed with and criticized by Navin Rai, Suzy Allen and Tom Headland. Their contribution has been invaluable. Jude Wong, of the University of Hawaii, Department of Linguistics, must be acknowledged for her computer wizardry. Ev Wingert is particularly thanked for his assistance with the maps.

I wish to thank Karl L. Hutterer, University of Michigan, Museum of Anthropology, who has for some time encouraged me to publish this material. Without this encouragement it is doubtful whether I would have returned to this work to prepare it for publication. At the University of Michigan Center for South and Southeast Asian Studies I would like to thank Jan Opdyke for her invaluable assistance and advice on many technical matters of publication, and Linda Deno, for typing the manuscript. Thanks also to Rachael Cohen for helping to unlock the mysteries of Word, and to David Akin for proofreading the final draft.

No amount of thanks can express the gratitude I feel for the Ayta families who hosted me and tolerated my many questions. Pan-Biri Paanan, Pan-Pungpung Paanan, Pan-Titiwən Paanan, Pan-Saligao, Pan Daytəm, Kapitan Pan-Aromi Mendosa, Kapitan Pan-Melicia, Pan-Boransa, Pan-Besil and all their families are to be especially thanked. During my time with them they fed me and sheltered me, and taught me the true definition of hospitality.

CHAPTER 1

INTRODUCTION

A. The Study of Anthropogenic Ecosystems

The present study is an attempt to assess the historical sequence of changing ecosystemic relations in upland Zambales from the perspective of succession theory. In so doing, I hope to demonstrate in a more general way how the concept of succession may be employed in understanding dynamic ecosystems in which humans are major actors.

The area under consideration lies within the Cabusilan Mountains in southern Zambales Province, Luzon, Philippines, mostly within the municipality of San Marcelino. These mountains are inhabited by a large Negrito population, who refer to themselves as Ayta. What is important for the present study is the fact that this area has been conspicuously deforested. Large areas of grassland and numerous diffuse patches of successional vegetation today dominate the landscape. Yet the present Ayta population has been able to maintain a high degree of self-sufficiency through a very productive system of swidden agriculture. Fox (1953:184-85) accurately noted that:

> the environment of the Pinatubo Negritos is not an immutable, natural phenomenon but is, in part, man made, and has been altered considerably. . . . the present threshold of Negrito culture is the product of many environmental adjustments.

In short, the ecosystem in which contemporary Ayta exist is a dynamic one in which there has been a long historical sequence of

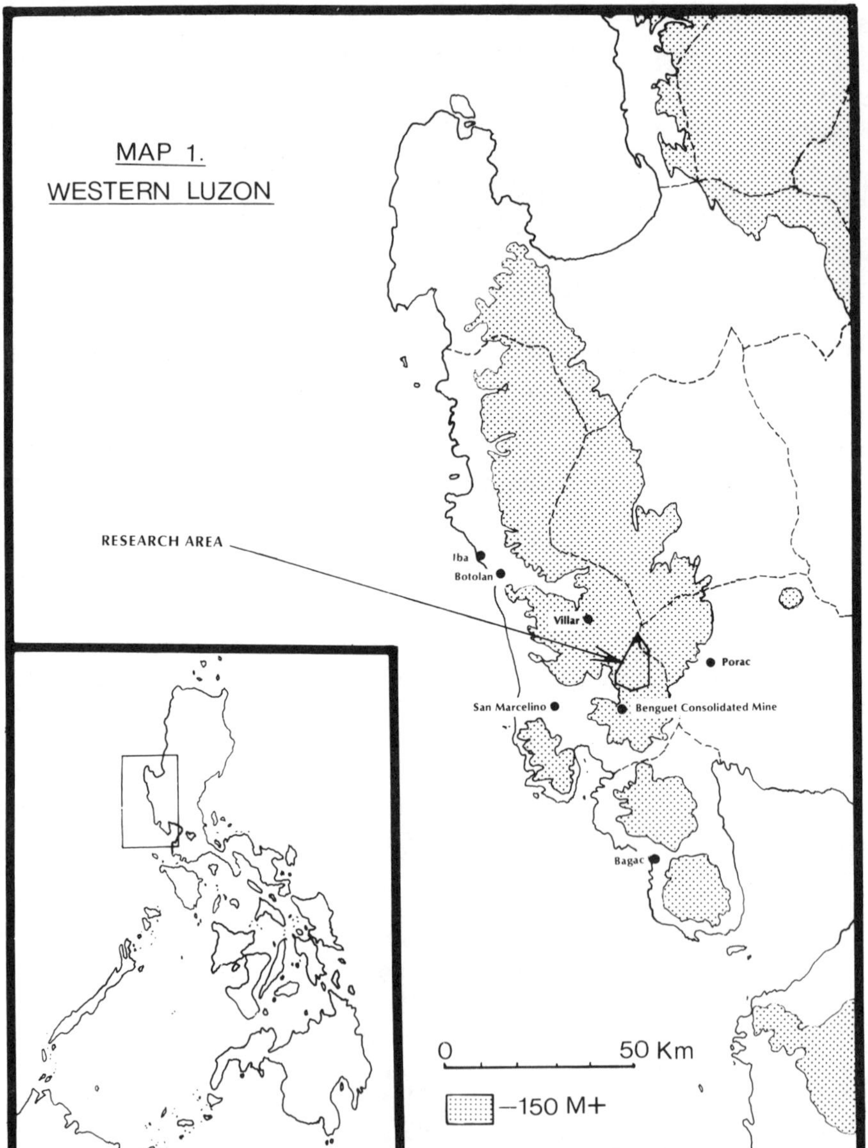
MAP 1.
WESTERN LUZON
RESEARCH AREA
Iba
Botolan
Villar
Porac
San Marcelino
Benguet Consolidated Mine
Bagac
0
50 Km
–150 M+

changing ecosystemic relations. Such a scenario raises several questions. The first concerns the precise nature of ecological change: what factors have caused and maintain deforestation and the resulting environmental patchiness characteristic of the area? Another question concerns the time frame of deforestation: how long has this been occurring and to what extent have the different causes varied in their degree and impact through time? Given that significant forest clearance has occurred, what sorts of successional associations have replaced them? And what circumstances determine the trajectory of specific successional sequences? Finally, given that various types of disturbances continue to occur, is it possible to make any statements as to the stability of the present environment?

The above set of questions are relatively straightforward and not particularly abstract. But they derive from, and reflect, an attempt to more parsimoniously view the ways in which human actors participate in the formation and maintenance of dynamic ecosystems by more explicitly incorporating the generative potential of succession theory into analyses of human adaptation. A fundamental point which I wish to make in this study is that the history of an ecosystem is more than a series of events which can only be described in the particular.

One of the goals in the development of ecological theory in anthropology has been the desire to conceive a theory of human adaptation independent of historical bias. Thus Hatch (1973:223-24, cited in Winterhalder 1980:140) voices "the desire to root institutions in a substratum of forces or processes that are immune to the vicissitudes of history." Such a goal has given rise to the mistaken notion that to do so one must deny the special role of a systems' unique history. As Winterhalder (1980:136) notes:

> Ecological adaptations result from historical processes in natural ecosystems which have as their most important characteristics temporal variance and spatial heterogeneity. Anthropological studies . . . give little attention to either, or to the historical qualities of natural environments.

Furthermore, he notes (1980:157):

> Much of what ecological anthropologists would like to discover--how human ecological systems reach or depart from stable configurations . . . --is lost or severely constrained if analysis dispenses with the study of historical routes of adjustment.

More succinctly, Richerson (1977:7) states, "As the organisms in a community evolve, many fortuitous events influence the course of evolution--natural communities have a history." In the case of anthropogenic ecosystems, understanding historical factors is perhaps even more important. The effects of human interference have been superimposed on top of, and absorbed into, natural ecosystems, resulting in systemic transformations over much accelerated time frames. The history of such systems is frequently accessible to us, though perhaps only through fragmentary or circumstantial evidence, and the sequence of events which has shaped a particular ecosystem can be understood. A theory of sufficient explanatory breadth should be able to account for the historical uniqueness of an ecosystem while at the same time avoiding a particularistic perspective. By placing historically unique events into a framework of succession theory, this is possible. This study is intended as a demonstration of how this might be done.

There are two other concerns informing the particular successional approach I propose here. The first is that ecological anthropologists have tended to place undue emphasis on homeostasis. Salisbury (1975:128) notes that:

> The ecosystem models most commonly used to organize the data obtained from traditional subsistence-based societies have been equilibrium models. . . . Much "ecology" has actually been modern neo-functionalism, explaining why traditional society was unchanging and/or "adapted to its environment."

The neofunctionalist program has certainly not gone unchallenged (Bennet 1975, Winterhalder 1980, Salisbury 1975). Although it is less prevalent today, one must still remain cognizant of and thereby avoid it. A second concern is in the way environments are characterized in studies of human adaptation. According to Winterhalder (1980:136):

> the descriptions which accompany most ecological anthropology analyses are narrative descriptions of common sense ecological features; the environment is presented as a static background characterized by averaging statistics.

As he further points out (Ibid.:136), "it is deviations from static, normative patterns that engender evolutionarily significant events." He stresses "the importance of variability and range, as opposed to means, in the analysis of human ecological adaptation and evolution."

In our attempts to understand ecosystems in which humans are dynamic actors, such as that which exists in upland Zambales, we require a theory which is explicitly ecosystemic and which can take account of the unique history of a particular landscape. Succession theory is a particularly appropriate model with which to do this. It provides not only a conceptual model, but an operational research agenda as well. The sorts of information required for a convincing successional argument are clearly specifiable.

Having defined what I believe the issues in this study to be, it is now necessary to provide some background on succession theory. In the following I will review the development of succession theory and discuss some of the assumptions upon which it is based. I will then examine what ecologists have identified as the main features of the successional process and, finally, I will consider some of the conceptual problems that exist with succession theory. As this discussion proceeds I would ask that readers keep one question, posed by Colinvaux (1978:123), in mind:

> Is there some . . . organization beyond the species level that fits communities together? Or [can] . . . everything . . . be explained . . . by an hypothesis that lets each species in the plant community act selfishly in the pursuit of its own Darwinian fitness?

B. Succession Theory

The concept of succession was developed primarily by botanists. According to Colinvaux (1978:121), it was the first unifying theme in ecology. The idea was first roughly enunciated in the nineteenth century, by von Humboldt (1805) and de Candolle (1855). Clements (1916) was the first to develop a unified theory and his formulation remains largely intact today. Clements recognized successional stages on floristic, structural and habitat characteristics, and focused attention on community properties, i.e., the inter-relation between plants and other factors. Communities, he stressed, must be regarded as whole entities in their own right. The major shortcoming of Clements' explication of succession was his use of the organismic analogy: this has subsequently been the root of a great deal of abuse of the succession concept. Clements outlined five stages of succession. These were:

1. Nudation: exposure of new substrate
2. Migration: arrival of disseminules
3. Excesis: germination, growth, establishment, reproduction
4. Competition: species replacement
5. Reaction: habitat change caused by invading species

Through time the concept of succession has been elaborated and refined, and a large number of definitions of the process have been given. Perhaps the best way to convey the major outlines of this concept is to review a number of these definitions:

> Tansley (1920:136) - "The gradual change which occurs in the vegetation of a given area of the earth's surface on which one population succeeds the other."

> Margalef (1968:27) - "the occupation of an area by organisms involved in an incessant process of action and reaction which in time results in changes in both the environment and the community, both undergoing continuous reciprocal influence and adjustment."

Odum (1971:251) - Odum's definition is rather long and will not be repeated here. This definition can be summed up as a process of "ecosystem development," a biological Markov chain of sorts in which every step is dependent upon those preceding it and in which, in the process, systemic relations develop.

Horn (1974:25) - "a pattern of changes in specific composition of a community after a radical disturbance or after the opening of a new patch in the physical environment for colonization by plants and animals."

Mueller-Dombois and Ellenberg (1974:371) - "Succession as a study orientation is concerned with all changes in vegetation that occur in the same habitat in the course of time."

Hardesty (1977:95) - "the process of change that takes place in an ecological community after colonization of a barren habitat."

Succession is more than simply an arrival sequence, or what Egler (1954) has called "relay floristics." Margalef states that "ecologists agree that succession follows a direction but opinions differ when a working definition, or even some criteria or common characters of the usual trend, are to be formulated in a concrete way." The UNESCO MAB manual (Slatyer 1976:9) outlines two key assumptions regarding the process:

(1) Species replacement during succession occurs because populations tend to modify the environment, making conditions less favorable for their own persistence and leading to progressive substitutions.

(2) Climax is the end point; the community is self-perpetuating and in equilibrium.

A successional pattern (species present, relative abundance through time) is determined by the product of two possibilities: (1) Propagules being available, and (2) being able to complete their life cycle. Successful colonization by available propagules depends on each particular species' adaptation to physiochemical site characteristics.

Several changes occur in the process of succession. These, which have been noted by Odum (1971), Margalef (1968), and Horn (1974) among others, include:

(1) Increase in biomass: Both primary production[1] and biomass increase, but the ratio of primary production to biomass decreases.

(2) Diversity increases: Species enrichment occurs throughout the process. Several factors influence species diversity: (1) modification of soils and microclimate, (2) increasing interaction between plants, (3) declining and/or changing frequency of seedling establishment, and (4) increasing stratification of the vegetation.

(3) Increase in proportion of inert or dead matter with low respiratory rate: Accumulation of organic materials. Annual production exceeds annual respiration during succession.

(4) Multiplication of niches: Longer food chains, more strict specialization.

(5) Plants--reduction in number of offspring and better production: r-selection vs. K-selection. r-species/strategies are opportunistic, present in the early stages of succession. They are annual and "in the business of living where accident has cleared away competition" (Colinvaux 1978:125). They are poor competitors, depending instead on wide, massive seed dispersal. Rapid growth and reproduction is the goal. K-species employ equilibrium strategies and are present in later stages of succession. They are perennials which spend "some coin on . . . food reserves . . . [and] on holding off the competition. An equilibrium species plans for babies in the

---

1. Primary production is "the amount of solar energy plants are able to capture as chemical energy" (Pianka 1978:47).

future rather than babies now" (Ibid.:126). More energy is put into supporting structures at the expense of leaf area and reproduction, and more calories are required to maintain them. They have an increased tolerance to shade and other stresses, have resistant root systems and are effective competitors.

(6) Decreasing ecological perturbation with establishment of cycles: Closing and tightening of nutrient cycles. Nutrients become sequestered in biomass, resulting in lowered gross productivity.

(7) Increased structural complexity.

As with any important theoretical formulation, despite attempts at precision in definition there have been problems in applying the idea of succession. These problems are both of a conceptual nature and of a more empirical nature, as succession is used in the interpretation of real data. As Horn (1974) points out, there is no single, unitary concept of succession. Rather, the concept is potentially full of ambiguities. With regard to the specific goals of the present study, three problems are of particular concern. These are: (1) teleological assumptions, (2) the nature of climax communities, and (3) the problem of scale and boundaries.

One of the single most serious problems with the concept of succession has been with regard to what Richerson (1977:3, 4) has referred to as the "fallacy of misplaced teleology." According to Richerson this "occurs when purpose is attributed to a unit of organization on which no creative process is known to operate" (Ibid.:4). This problem becomes particularly acute when relating ecological concepts to human affairs. Biological principles or systems in which emergent properties are perceived seem, incorrectly, tailor fit to anthropological needs. Richerson's dictum (Ibid.:4) is therefore of critical importance:

> In fact no teleological theories appear to be required except those operating at rather low levels of organization--individuals and groups of closely related individuals--in either genetic or culture-based systems.

With regard to succession, the roots of teleological thinking can be traced back to Clements' use of the organismic analogy in his explication of the concept. Examples of this kind of thinking are plentiful, and terms such as "directed change," "recovery" and "strategy," implying some sort of goal-orientation, fill the succession literature. Odum, for instance, reverts to the organismic analogy by asking the "intriguing question: Do mature ecosystems age as organisms do?" (1971:257). The problem is made all the more troublesome by the fact that teleological assumptions are often not overt but subtle. Such teleological thinking seems to emerge as the result of two factors: (1) the predictability of successions, and (2) the recognition of certain community-level properties. A more productive way of looking at succession is simply that it should be considered only insofar as it is consistent with basic Darwinian principles. Colinvaux (1978:127) notes that "The apparent community properties are simply explained as a consequence of numbers of Darwinian species each doing its own thing."

With regard to the nature of climax communities, it must be noted that the climax concept is one of the most controversial aspects of succession theory. A wide range of definitions has been offered for it. Margalef (1968:32) sees climax as the final, equilibrium stage of succession. Pianka (1978:62) defines it as "A stable community in dynamic equilibrium that replaces itself": production equals respiration. Finally, Mueller-Dombois and Ellenberg (1974:397) state, "A succession ends with a community in which the species perpetuate themselves through reproduction." Species composition and structure are stable over long periods of time. Horn, among others, notes that climax has been idealized as a state of "constant specific composition" (1974:25), implying a nondynamic system, when it is actually a "blurred successional patchwork" (Ibid.:30). Most ecologists currently accept the view that climax is a state of dynamic equilibrium. One issue which persists in discussions of climax concerns nonterrestrial or cyclically disturbed ecosystems. For instance, freshwater stream environments are undergoing constant replacement and are not generally considered to be in a climax state. Yet even speciation can occur in such environments, implying long-term stability. Another case in point is, in the tropics, areas such as western Luzon which are covered by *Imperata* grass. Such associations, once established, are generally permanent, being burned annually. Successful reproduction even relies on this. Yet these areas are generally not considered to be climax communities. Any concept of climax, if it is to be at all useful, must take such cases into account.

A third problem concerns scale and boundaries. Margalef (1968:39) states that "ecotones often disappear when we look for them." This statement most aptly expresses the problem of boundaries in succession theory, a problem which has not been sufficiently addressed and is far from resolved. Ecotonal species are "independently assorted," making boundaries invisible without detailed vegetation ecology procedures. Even then such boundaries may only be statistical entities.

C. Context and Disturbance

Mayr has asked "how can . . . environmental qualities be described so that they articulate with the theory necessary to account for specific adaptive forms. . . ." (Mayr 1976:103, quoted in Winterhalder 1980:136). I have proposed succession theory as an appropriate model for explaining the sequence of changing ecosystemic relations in upland Zambales. Succession theory imposes an agenda which is explicitly ecosystemic, while at the same time requiring that the circumstances impinging on a given successional sequence be specified. It compels us to define the interactions between elements throughout the process of ecosystem development. Equally important, succession theory suggests a framework for defining the types of data to be collected in an actual research program, and provides a conceptual framework for interpreting these data. It remains to specify precisely what types of data are necessary to provide a convincing successional argument.

The attempt to understand the long-term dynamics of particular anthropogenic ecosystems from a successional perspective requires specification of two classes of data. These are (1) context and (2) disturbance.

In speaking of "context" I am referring simply to the physical stage upon which successional processes in a particular setting occur. More specifically, the contextual components of a particular landscape include the physiographic, climatic and biotic features which characterize that area. It requires no explanation to see how, in a general way, these factors are of critical importance in determining the successional trajectories characteristic of a particular area. Each class itself, however, may influence successional processes in complex ways. Chapters 3, 4 and 5 are concerned with specifying the particular contextual characteristics of the Cabusilan area and with defining how they may act to determine particular successional pathways.

The second part of this equation is what sets a new successional sequence in motion: the disturbance. By disturbance is meant an event which alters the biotic components of a specified area. Grime (1979:7) states that "disturbance is associated with the partial or total destruction of the plant biomass." He further notes (Ibid.:39) that "Forms of disturbance differ with respect to their selectivity. . . . In general, there is an inverse correlation between degree of selectivity and intensity of disturbance." Farnworth and Golley (1973:118) note that "Succession is initiated by disturbance that substantially destroys the structure of the . . . community." Winterhalder (1980:154) clarifies this further, stating "From a localized viewpoint the disturbance-succession cycle is one of abrupt followed by slower change in multiple elements of habitat, including productivity and species composition." Slatyer (1976: preface) states that:

> There exists a wide range in the impact of man's actions on natural ecosystems. When impact is limited, the ecosystem tends to retain most of its original attributes. When impact is very severe, the natural system is often, in effect, replaced.

Attention to the disturbance factor is not new. It was the recognition of human impact on the Nebraska prairie which stimulated Clements' definition of the stages of succession. Gams (1918) was the first to point specifically to the importance of specifying the disturbance which engenders a new successional sequence. As Mueller-Dombois and Ellenberg (1974:372) point out:

> Gams (1918) made the point that it was unnecessary to distinguish kinds of successions by whether they originated from natural or man-made disturbances. . . . Instead Gams emphasized that successions differed

> fundamentally by the pattern of timing of the perturbations.

On this basis he distinguished between "normal" sequences, "rhythmic" sequences, and "catastrophic" sequences. Tansley (1939) also recognized the significance of disturbance when he distinguished between "autogenic" and "allogenic" successions. It is allogenic succession, i.e., "vegetation change due to site changes" (Mueller-Dombois and Ellenberg 1974:375) which is relevant to our interest in disturbance. Richards (1952:378) was more explicit in his recognition of the significance of disturbance, and much of his discussion of succession is framed in terms of disturbance factors.

A focus on disturbance is useful for a number of reasons, many of them coincident with a more general successional approach. First, it takes account of a particular system's history. Second, such an approach is explicitly nonhomeostatic, thereby avoiding the sort of ecological functionalism discussed previously. The dynamism of an ecosystem, and the human response to that dynamism, can be clearly specified. Most significant perhaps is the predictive potential of this approach. One problem that has habitually plagued ecologists, particularly those working in the tropics, is the variation in specific successional seres[2] within a single area. Mueller-Dombois and Ellenberg (1974:396) state that:

> Certainly, not all variation in the vegetation of a region can be attributed to intrinsic environmental variations between habitats. On the same topographic habitat type, one may find several different plant communities in the floristic region. Such differences are primarily caused by partial disturbances resulting in secondary successions.

If we can understand why humans choose to disturb specific areas, how and to what degree they do this, what sort of succession results and how they respond to specific successions and to the aggregate successional mosaic, we will have succeeded in realizing some of the

2. A "sere" is a successional sequence, i.e., a temporal sequence of changing plant species in a given area.

predictive potential of the successional approach. It remains to demonstrate more specifically how this is done.

Slatyer (1976:12) states that "before the stability of any real community can be discussed, three scales must be specified: the frequency, the area and the intensity of perturbations." It is by specifying the parameters of disturbance--frequency, scale and intensity--that we are able to begin understanding those factors promoting both stability and dynamism in particular ecosystems.

Gams (1918) was the first to call attention to the significance of the frequency of disturbance. Much more recently Pianka (1978:62) has stated:

> Disturbances, both man made and natural . . . are often frequent enough that extensive areas have not had time enough to reach their own climax state. An equilibrium is reached whereby the proportion of a habitat supporting early successional stages is determined by the frequency of disturbance.

Farnworth and Golley (1974:119) are also cognizant of this factor when they state:

> Repeated disturbance results in destruction of disseminules present in the soil, as well as the reduction of the ability of the soil to support plant growth. Repeated disturbance commonly causes a deflection from the primary community end-point toward another community type.

The scale or area of a given disturbance is an equally significant factor in determining the course of particular successional sequences. It has received more attention in the literature than either frequency or intensity. Succession in a small plot surrounded by forest will be very different from that in a much larger area. Slatyer (1976:12), for instance, notes that "Because the area is small, seeds both of nearby low-vagility plants and of more distant high-vagility species will colonize the gap." Farnworth and Golley (1974:119), on the other hand, remark that "On very large areas, the invasion of disseminules must necessarily take place from the edge of the plot so that full stockings of the stand may be greatly delayed." Opler et al. (1977:395) note that:

> This distance factor will be of increasingly critical importance in the tropics. As the ratio of disturbed areas to climax has increased, instead of small cultivated islands surrounded by forest, the historical tropical land use technique, small forest islands are being surrounded by vast expanses of perturbed areas . . . . In these circumstances, the potential rate of ultimate natural recovery is being constantly lessened.

The effect of intensity of disturbance on successional potential has received less explicit attention, and its significance has been underestimated. The intensity of a disturbance determines, among other things, the degree of selective stress imposed on component populations of a community, and thus determines which components survive to begin the succession process. As Slatyer (1976:12) notes, "if the disturbance is severe . . . recruitment must come from outside. . . ." Grime (1979:41) states that:

> low intensities of disturbance function as a modifier of competition by favoring those species which tend to maintain their competitive ability either by avoidance of damage or by rapid recovery from its effects. However, when we turn to productive habitats exposed to repeated and severe disturbance, it is apparent that competitors are excluded and that a quite different strategy . . . prevails.

A grassland fire is of very different intensity than one in secondary scrub forest, and a swidden cleared from such forest burns differently than one cleared from a wild banana grove. Likewise any fire which burns in an area subject to a long dry season will likely be different from one in an area with evenly distributed rainfall. There is clearly a range of parameters of disturbance intensity.

Most of the observable disturbances in upland Zambales have been initiated by humans, and it is the historical sequence of these disturbances which accounts for the present mosaic environment within the Cabusilan area. These anthropogenic disturbances will therefore be the object of focus. This is not to deny that natural disturbances are not important. Typhoons, landslides and tree-falls have always occurred. But the impact of these has been localized and, on a regional scale, insignificant. Only with the presence of humans has large-scale ecological change occurred.

In upland Zambales, three types of disturbance have been significant in producing the current ecological patchiness characteristic of the area. These are (1) swiddening, (2) burning, primarily of grasslands, and (3) logging. In Chapters 6, 7 and 8, respectively, the parameters of disturbance for each will be considered. These three have existed in an historical sequence of varying importance and effect through time. By defining this changing sequence of disturbance we may thereby come to better understand the historical range of adaptive processes both initiated by, and experienced by, the human actors who have created the contemporary upland agroecosystem in Zambales. The current pattern of disturbances is indicative both of present adaptive strategies and of processes which may occur in the future.

The creation and maintenance of environmental diversity in the Cabusilan area is the result of several types of human activity, which have as their central feature disturbance of some portion of the upland ecosystem, whether previously disturbed or not. In what follows I attempt to apply the successional methodology to this case. In so doing I would hope to demonstrate in an empirical way the utility and generative potential of this body of theory. The application of a successional methodology requires specification of certain types of well-defined information. After providing a brief description of Ayta subsistence and settlement in Chapter 2, I turn to a discussion of contextual factors. Chapters 3, 4 and 5 consider physiography, climate and vegetation respectively. The chapters following this--6, 7 and 8--will examine the three primary types of disturbance which have occurred in the Cabusilan area: swiddening, burning and logging. The parameters of disturbance for each--frequency, scale and intensity--will be considered, specifically in terms of the type of impact each may be expected to have. In Chapter 9 an attempt at synthesis will be made and the sustainability of the present system will be assessed.

A major goal of this discussion will be to explicate the historical sequence of changing ecosystemic relations which have occurred in upland Zambales. In attempting to understand any system, we must be cognizant of its genesis and formation, and must also recognize that that system continues to change and evolve. If the principles determining this can be clearly specified, a measure of predictability may be achieved. The predictability of specific successional sequences, based on lower-level ecological principles, is what makes the anthropological applications of succession theory so attractive.

CHAPTER 2

AYTA SUBSISTENCE, SETTLEMENT AND SOCIAL ORGANIZATION

A. Background to Ayta Subsistence and Settlement

Contemporary Ayta adaptive strategies cannot be understood from a purely synchronic frame of reference. Rather, they are the result of a long temporal sequence during which major ecological, demographic and economic changes have undoubtedly occurred. The reciprocal relationship between human actions and environmental change in upland Zambales has been accompanied, and in one sense driven, by a change in subsistence strategies. Specifically, it appears that, through time, Ayta have relied progressively less on hunting and gathering, and more on agriculture.

The evidence that hunting and gathering was once the dominant mode of Ayta subsistence is rather circumstantial, and I must underscore the speculative nature of the present discussion. Nevertheless, the evidence is sufficiently compelling to merit discussion. I believe it likely that at one time, sometime prior to Spanish contact, the Ayta were exclusively hunter-gatherers. Presumably they might have been something like contemporary Agta in northeastern Luzon (Estioko-Griffin and Griffin 1975, 1981; Peterson 1978; Rai 1982). If such were the case, heavy reliance would have been placed upon the hunting of pig and deer, fishing, and the procurement of carbohydrates from sources such as wild yams (*Dioscorea*) and the *Caryota* palm. Trade for agricultural products would presumably also have been an important component of this adaptation. A scenario such as this is consistent with the view of Fox (1953:249-50), though he stresses the greater relative importance of plant foods over meat during a presumed preagricultural period. Fox also believes that the

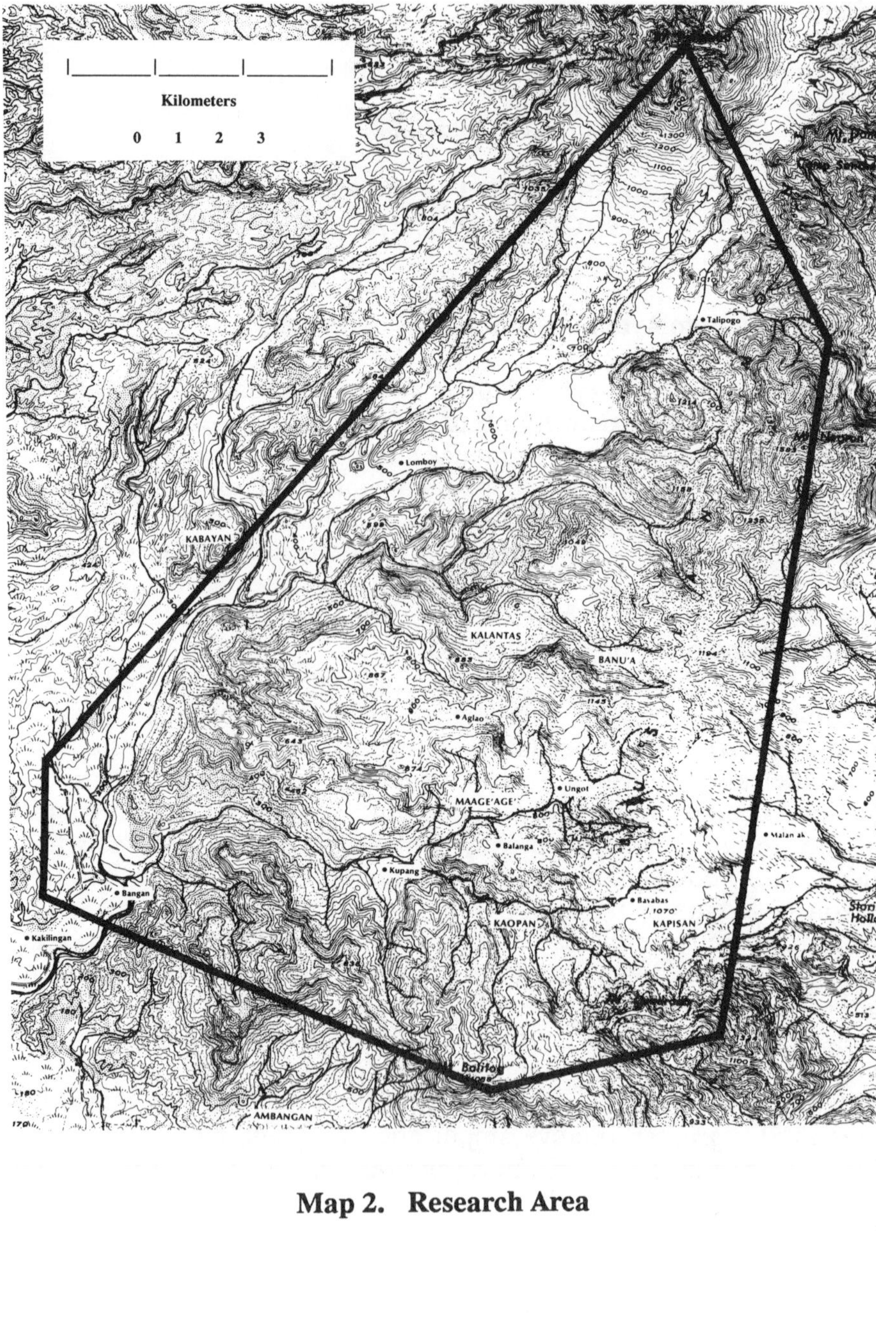

Map 2. Research Area

Ayta were never forest dwellers, but have only been driven into more forested interior areas by pressure from lowland populations (Ibid.:250).

Among the more compelling pieces of evidence that Ayta were once hunter-gatherers are oral accounts. Ayta claim that according to their grandfathers, the first people (*onan tao'*), their ancestors, were *amukok*. *Amukok* are characterized as having no agriculture, either lacking clothes or wearing bark cloth, not building houses, living in rock-shelters, and in general being very "wild." While showing informants an article on the Tasaday of Mindanao, the response was that these were indeed "real" *amukok*. Certain wild food plants, such as the Zingiber *Amomum sp.*, were referred to as *kanən amukok* (*amukok* food). On other occasions children were observed playing *amukok* by making loincloths out of wild banana sheaths. I think it is significant that when Ayta speak of *amukok*, it is in the sense that these existed in the far distant past rather than within historical time. Since carrying out my research in Zambales, I have worked with Penan hunter-gatherers in Sarawak. Groups of Penan such as the Penan Suai and Penan Jelalong, who have been settled for perhaps two hundred years, still recall direct genealogical links with ancestors who were nomadic. They speak of past generations when their ancestors were nomadic as within historical time, not in some vague, distant past, as among Ayta. This would seem to underscore the very long time that Ayta have practiced swidden agriculture.[3]

In addition to oral accounts, there is also the indirect evidence of viewing Ayta relative to other Philippine Negrito groups, nearly all of which in the historical or ethnographic record have been hunter-gatherers. Such evidence is certainly not conclusive, but does lend support to Ayta oral historical accounts. Together they suggest that at sometime in the past the Ayta were predominantly hunter-gatherers.

Ayta involvement in trade with lowland groups has certainly been occurring for many centuries, if not millennia. In 1947 Fox located two sites, Purakin and Alibaiyan, near the present day Ayta village of Ugik, to the north-west of the research area. These sites

---

3. Ayta in fact speak of their *amukok* ancestors in much the same vague way as Kenyah in Sarawak speak of a past reliance on hunting and gathering. Kenyah, if they ever were hunter-gatherers, have certainly been accomplished agriculturalists for many centuries.

yielded late Sung, Yuan and early Ming ceramic fragments (Evangelista and Madale 1961). In 1961 Evangelista excavated two sites, Nahing and Gandong, in the same vicinity (Ibid.). These sites yielded twelfth to fourteenth century porcelains, earthenware and stoneware. The presence of such goods in the uplands clearly indicates that trade was occurring. If we can extrapolate from the example of other Southeast Asian hunter-gatherers and from sparse historical accounts, primary Ayta exports would have included items such as rattan, beeswax, honey, medicinal herbs and, it appears, deer skins. Trade relationships were likely institutionalized, much as is the case among the Agta of northeastern Luzon today (Peterson 1978), or they may have been considerably more fluid. In the late seventeenth century, in their attempts to settle inland populations, the Spanish tried to institute a trade prohibition between upland and lowland groups "in order that, being deprived of that recourse, hunger or necessity might compel them to descend from the mountain and live in a settlement in order to exist" (Blair and Robertson 1909, 43:42). Coastal groups apparently had a very strong interest in maintaining this trade, a situation which effectively countered Spanish attempts to move these groups out of the mountains. Gemelli Careri (1963:34) writing in 1696 reported that:

> The Christian Indians, that they may have all the profit of the wax the blacks gather in the woods, have persuaded them by all means to shun coming under the Spanish yoke, because they would be obliged to pay tribute.

Trade was likely institutionalized to the point that Chinese middlemen were involved. After the Chinese insurrection of 1639-1640, the bodies of six hundred Chinese were encountered in coastal areas of Zambales by Spanish patrols (Blair and Robertson 1909, 27:227). Gemelli Careri (1963:33) speaking of the upland Sambal stated, "True it is, that by means of heathen Chineses [*sic*], who deal with them in the mountains, some deformed statues have been found in their huts."

When Ayta began to practice swidden agriculture cannot be precisely known. Swiddening appears to have become established in the area very long ago, possibly before the Spanish colonial era.

Based upon historical accounts, the present degree of environmental degradation, the current large number of cultigen varieties, and current population densities, the initial adoption of swidden agriculture by Ayta cannot have been a recent phenomenon.

The initial expansion of swiddening activity presumably began under large scale forest cover, with an Old World, Southeast Asian crop complex. This would have included crops such as taro (*Colocasia esculenta*), yams (*Dioscorea alata*), bananas (*Musa paradisiaca, M. sapientum*), rice (*Oryza sativa*) and numerous minor cultigens. The procurement of nondomesticated carbohydrates such as numerous species of *Dioscorea*, *Caryota cumingii*, and *Amorphophallus* would likely have continued. Hunting during this period was probably of great importance and game was certainly more plentiful than today. Extrapolating from what we know of other contemporary Philippine Negrito groups, pigs and deer would have been primary sources of animal protein, along with fish and eels.

It is likely that the introduction of New World crops, notably sweet potato (*Ipomoea batatas*), *Xanthosoma*, corn (*Zea mays*) and cassava (*Manihot esculenta*), brought about a suite of ecological, demographic and economic changes. The most important effect would have been an expansion of the swidden system. Ayta subsistence is today entirely dependent upon carbohydrates derived from swiddening; New World crops hold an overwhelmingly dominant position in this system.

B. Contemporary Ayta Subsistence

For contemporary Ayta, swidden agriculture today provides by far the greater part of their nutritional needs. The hunting of larger types of game such as pig and deer is negligible, and even smaller types of game are consumed in very small quantities. As I will discuss swiddening in detail in Chapter 6, the present section will focus on other aspects of contemporary Ayta subsistence, specifically hunting and fishing.

The Ayta have a sophisticated knowledge of all aspects of their environment. Fox (1953:187) speaks of "The acute observation of the pygmies and their awareness of the interrelationships between plant and animal life giving them an ecological picture of their environment," and states (Ibid.:188) that "Most Negrito men can with ease enumerate the specific or descriptive names of at least 450 plants, 75 birds, most of the snakes, fish, insects, and animals, and of even 20 species of ants." As an example Fox notes that many plants are known in the context of their relationship to other ecological factors, such as the birds which feed on them. On one occasion informants I spoke to were able to enumerate 43 species of plants eaten by birds, 17 species by palm civet, 38 by fruit bat, 12 by pig and 21 by deer. Their locations, seasonal incidence of occurrence, and other such features are a major component of this large body of specialized knowledge.

In the first part of this chapter I discussed the apparent expansion of swidden agriculture in the area over a period of several centuries. Accompanying this process of swidden expansion has been a marked depletion in the availability of all types of game. The extent to which game, both large and small, has been depleted in the Cabusilan area is striking. Pig and deer have disappeared almost completely, remaining only on the upper north and east slopes of Mt. Pinatubo. Informants state that two pigs remain on Mt. Cuadrado (see Map 2). Smaller types of game have been seriously depleted as well. Philippine palm civet (*lamidan*; *Paradoxorus philipensis*), the largest animal still hunted and trapped, is taken only very infrequently. In eight months of field research I saw palm civet consumed only once. Monkeys (*sakə'*; *Macacus sp.*), though still said to exist in the area, were never seen in the wild or successfully hunted during my time there. Fruit bats (*panilao*; *Pteronidae*) are uncommon as well. Only two were shot during the period of research. The greatest percentage of animal protein is derived from two sources--birds and river resources.

Avian resources comprise about 50 percent by weight of the animal protein consumed. Large game birds such as the wild chicken (*manok dikət*; *Gallus gallus*) and hawks (*lawin*; *Haliastur indicus*) are no longer plentiful. During eight months of fieldwork the largest bird taken was an owl (*bu'ok*; *Bubonidae*), weighing .4 kilograms. Large hawks were spotted on occasion, but never successfully hunted. The hornbill is extinct in this area.

Though large birds have nearly disappeared, the area abounds in numerous small birds, whose habits are very well understood by the Ayta. In excess of eighty such species of birds are recognized. In one instance informants were able to enumerate sixty-five species, and provide descriptions of habitat, food preference, song and idiosyncracies of behavior for each.

The hunting of birds is carried out by men and boys. The bird arrow, generally called *balangat*, is always used. Many subtypes exist, but all are characterized by having three to five notched or barbed points, with a shaft from 4½' to 6' in length. Hunting occurs during the day or, with a flashlight, at night. Today some airguns are also being used.

Riverine resources also provide a large percentage of the animal protein consumed by Ayta. River shrimp (*ulang*; *Leander sp.*, *Macrobrachium sp.*), small goby-like fish (*Pyhacichthys aspro*) and occasionally frogs (*pahinga*; *Rana sp.*) are the primary riverine species procured for consumption. Only on rare occasions are eels taken and these tend to be quite small; the largest I recorded was .3 kilograms, and this was exceptionally large. This is in contrast to Northeastern Luzon, where the Agta frequently procure eels weighing 2 kilograms or more. Two primary methods are employed by Ayta in procuring shrimp, fish and frogs. Most important is *mangantoko*, in which goggles (*antoko*) and a small rubber-strap propelled spear (*asido*) are used to shoot fish and shrimp. According to Fox (1953:279) this method was introduced in about 1925 and "has completely revolutionized their fishing activities." A variation on this is night fishing (*mangatang*; from *atang*, lamp) with the aid of a flashlight. This is only done in shallow water, where fish, shrimp and frogs are clearly visible in the light. They are shot with the *asido*, or with a *balangat* to which a rubber strap is attached. It is possible to procure many large shrimp with the use of flashlights, and it is generally only by this method that frogs are taken in any quantity. The second Ayta method of fishing is by the building of small dams (*mamatiktik*). This method is generally only employed during the annual dry season, when water levels are low. At a spot where a stream splits into two a channel is cleared on one side and a dam built on the other. The resulting lower water level on the dammed side makes it possible to reach under rocks, into exposed leaf litter and extract shrimp. In most cases the labor expended in this activity is great in relation to the yield. In one case observed, two and a half hours were spent in this activity, during which large boulders, several over a meter across, were moved or channeled under. The result was approximately twenty small shrimp. Reed (1904:48) presents an account of this same method:

> In the small streams he builds dams of stones which he covers with banana leaves. Then with the bow and arrow he shoots the fish in the clear pool thus formed. Not infrequently the entire course of a creek will be changed. A dam is first made below in order to stop the passage of the fish, and after a time the stream is dammed at some point above in such a way as to change the current. Then, as the water slowly runs out of the part thus cut off, any fish remaining are easily caught.

With regard to terrestrial and avian game resources, most Ayta hunting is opportunistic rather than deliberate. That is, it takes place in the context of other activities or as the result of chance encounters, rather than as planned efforts. This is particularly true in the case of snakes, large lizards and birds, which are seen in the course of traveling from place to place or encountered while working in swiddens. The only type of game which is deliberately hunted is the palm civet (*lamidan*). These, however, are extremely uncommon and very wary; the few efforts made to actively hunt them are generally unsuccessful. The most reliable method of procuring these is by trapping during the wet season. Snares (*dalangdang*) are placed on logs which cross rain-swollen rivers. Trees may be purposely felled for this reason.

With regard to hunting, it must be noted that the Ayta possess what is certainly among the most sophisticated bow and arrow complexes possessed by any group in the Philippines. Fox states that "The Pinatubo pygmies distinguish more than fifty (!) types of arrows and have either specific or descriptive terms for them" (Ibid.:265). I was able to record twenty-six types during the period of research, still a very large complex. Some arrows have very specific uses, while others are generalized. There are arrows for pig and deer, birds, palm civet, and for other types of game. It is also important to point out that the bow and arrow is considered a necessary defensive weapon by Ayta, and there are a number of arrow types designed specifically for this purpose. All Ayta men carry a bow and arrows wherever they go.

A set almost always includes the multipointed *balangat*, in the event that birds are encountered, as well as several more generalized arrows carried exclusively for defensive purposes.

C. Causes of Resource Depletion

Terrestrial, avian and aquatic resources have all been seriously depleted in upland Zambales. Though the depletion of certain classes of game is relatively recent, most has occurred over a very much longer period of time. Several factors can be seen as contributing to the depletion of protein sources in the Zambales area. Most significant are the reciprocal factors of habitat change and increase in population density.

As noted, pig and deer have almost completely disappeared within the research area. This has been a recent phenomenon, since shortly after World War II. Ayta informants are unanimous in their agreement that such game animals existed until this time. Many commentators have also mentioned the presence of pig and deer in the not too distant past. Reed, for instance, notes (1904:44-45) that the Ayta environment "does not supply a great variety of game, but there are always deer and wild boars in abundance." Antonio Rodriguez, a wealthy Ilokano plantation owner who previously supervised logging activities in the Cabusilan area and whose family has had dealings with Ayta for several generations, stated that he himself had hunted pig on the western slopes of Mt. Balitog as a young man in the 1930s and 40s, and that his family used to give Ayta guns in exchange for pig meat.

As deforestation was occurring long before the present century, habitat change alone cannot be invoked as causing the depletion of game and other such resources in upland Zambales. In any case, both pig and deer are known to inhabit grassland and secondary forest areas elsewhere in the Philippines and Southeast Asia. Diong (1973:127), speaking of the Malayan wild pig, notes that "The most propitious combination of habitats for the wild pig is apparently forest or scrubland and cultivated fringe. . . ." Alcala (1976:145), referring to deer, noted that they are "herbivorous, feeding on leaves of shrubs and low trees as well as on cogon grass." Another source (Dept. of Nat. Resources, n.d.:[35]) notes that the deer:

> usually stays inside dense forests . . . during the day, but at night goes out to nearby cleared areas. It feeds close to forest edges, in the open grasslands, inside clearings, and even inside cultivated areas planted to corn or other crops that are located not far from forest patches.

That habitat change is not the cause of game depletion in the Cabusilan area is further evidenced by the fact that these species, particularly deer, are frequently hunted in lower elevation *Imperata*/scrubland areas in the vicinity of present-day Villar, a long-established Ayta settlement (War Office 1908:473) to the northwest of the research area. Habitat change, while it may have had some effect on relative abundance, cannot be invoked as causing the total depletion of these species within the research area. The only other alternative explanation is predation pressure. Such pressure is an artifact both of increased population density and of technological innovation. With regard to technological innovation, Fox (1953:262) notes that "in recent years many guns have been utilized by the pygmies, particularly in hunting wild pigs and deer. These were primarily locally made shotguns, called *paltik*." He further (Ibid.:262) states that:

> In 1945 during the liberation of Zambales, the pygmies obtained hundreds of rifles from Japanese soldiers who were killed while attempting to hide on the forested slopes of Mt. Pinatubo. However, these Japanese rifles will soon be useless, for already most of the ammunition has been consumed and it cannot be replaced.

This last prediction turned out to be inaccurate; the Ayta continued to use rifles until the 1972 imposition of martial law. The use of guns presumably increased the efficiency of hunting considerably, especially in an area subject to game drives by burning. Hames (1979:250) found in the Amazon that "the shotgun is a far superior hunting implement to the bow, and when the shotgun is coupled with other technological innovations such as headlamps . . . its efficiency increases further." He found hunting with guns to be 343 percent more efficient than hunting with bow and arrow.[4] I suggest that this best accounts for the rapid and nearly complete depletion of pig and deer within the research area.

The depletion of avian resources can be accounted for both by habitat change and by predation pressure. This has been more a change in species composition than depletion *per se*. Birds are still plentiful within the research area, but presumably of very different species composition than would have originally been the case. Primary rainforest, such as still exists in Northeast Luzon, supports large numbers of birds, many very large. These include hornbills, hawks, eagles, wild chickens, and many species of doves. These species (except for hornbills and, perhaps, eagles) still exist within the research area, but are extremely rare. The forest areas necessary for the reproduction and specialized feeding requirements of such species is not sufficiently extensive. Added to this has been predation pressure resulting both from Ayta population expansion and from the introduction of shotguns and airguns. Under these circumstances the population of birds characteristic of primary forest habitats have been greatly depleted. At the same time, the populations of numerous species of small birds characteristic of secondary habitats have increased. These include *Pycnonotus spp.*, *Oriolus chinensis*, *Microselis sp.* (Fox 1953:287) and numerous other species. While the primary forest habitat for endemic species has decreased, the habitat for these more generalized species characteristic of secondary vegetation has increased tremendously. This is in large part a function of the large number of prodigiously fruiting trees and herbs characteristic of secondary vegetation. Most notable are the many species of *Ficus* and the very abundant wild banana (*amukao*). Bunches of *amukao* fruit can be seen everywhere, at all stages of development. The fruits of *Ficus* are seasonal, but with the numerous species characteristic of the area, some fruiting appears to occur the entire year round. Gramnivorous bird species are also favored in such an environment. As a rule, those birds characteristic of secondary vegetation zones

4. It should be noted, however, that meat consumption of the group using shotguns was only 16 percent higher than the group using bow and arrow. Hames hypothesizes (Ibid.:250) that "time allocation to hunting is limited by hunting success (or efficiency), thus preventing overexploitation."

have much more generalized feeding habits and do not rely on a single food source, giving them a decided adaptive advantage. A highly productive secondary habitat has allowed the expansion of numerous small bird species.

The depletion of riverine resources is also due to habitat change and predation pressure. The conditions of the upland watershed have been altered considerably by deforestation. One effect of this deforestation has been the disappearance of many second and third order streams, resulting in decreased year-round drainage density and hence a reduction in habitat. This also presumably impedes the delivery of detritus and nutrients to the major streams which support most river life. The greater number of people presently living within the research area has also served to deplete river resources. This is amplified by the further habitat alteration resulting from the *mamatiktik* dam-building fishing method. The movement of rocks, the manipulation of water levels and the removal of leaf litter act to upset river microhabitat conditions. This type of fishing occurs throughout the dry season and certain suitable areas may be repeatedly disturbed.

The difference between fishing here and among the Agta of Northeast Luzon is remarkable. Among Ayta the average yield of one person fishing for one hour with spear and goggles is generally less than .01 kilogram. In Northeast Luzon the average for the same method is between .25 and 1.0 kilograms (Navin Rai, personal communication).

D. Settlement

Relative to other swidden groups in the Philippines, contemporary Ayta settlement patterns are extremely variable. As is the case in most Philippine swidden societies, Ayta settlements are generally dispersed (Conklin 1957; Frake 1960; Schlegel 1979; Wallace 1970). The extreme variability is observed in the size of settlements, length of occupation, composition, and in the types of shelters which are constructed .

Ayta settlements range in size from single isolated lean-tos, to rather large communities of five to fifteen or more thatched houses. The most frequently observed type of settlement are those consisting of a cluster of two or three huts. In a general way there is a relationship between the size of a settlement and how substantially it is constructed, though this is by no means absolute. Houses in these larger settlements tend to be more substantially built, and these settlements are more permanent than is generally the case. Lomboy, Talipogo, Kupang and Bayabas (see Map 2) are examples of such settlements. Two of these, Talipogo and Kupang, were only established during the year prior to my field research. Tima (personal communication) suggests that a regular process of fusion and dispersal is occurring, wherein such settlements, centered around a group of kindred, form, persist for several years and, due either to economic considerations or to conflicts, disperse. Such indeed appears to be the case, several such cycles having occurred in recent years.

Some of the variability we observe in Ayta settlement is a function of seasonality, but this does not account for the total range. Settlement does vary seasonally but not uniformly so. During the dry season, when households are clearing and planting new swidden fields, settlement is more dispersed. If swidden fields are at any distance from a settlement, families erect temporary shelters adjacent to their fields. These are generally in the form of *haong*, lean-tos constructed of *amukao* leaves. In some areas, notably Kalantas, families also inhabit rock-shelters. Though they may maintain houses in larger settlements, many families visit them only sporadically during the dry season, preferring to remain in the smaller dispersed settlements adjacent to their swiddens. Again, variability is the rule. For instance, the high elevation settlement of Bayabas, composed of six houses and situated near a clustering of swiddens, is almost never occupied by more than two or three families at a time. Most of these families maintain houses in Balanga or Kupang or are otherwise dispersed at different swidden sites. Given the steep terrain characteristic of the area, settlements are found over a wide elevational range, from as low as 250 meters to over 1000 meters. Most settlements are located between 500 and 1000 meters. Frequent moves between settlements at different elevations are common, particularly where a household has established swiddens in different locations. For instance, the inhabitants of Maagə'agə', between 300 and 500 meters, make frequent visits to Bayabas, at 940 meters, in order to procure swidden products.

E. Social Organization

Three levels of organization can be recognized in Ayta society: the nuclear family, the camp or hamlet, and the *sakop* or district. Most fundamental is the nuclear family, composed of husband, wife or wives and offspring. Each nuclear family functions as an economic unit, primarily by clearing and maintaining one or two swiddens per year, and may or may not occupy a separate dwelling. Nuclear families are normally associated with one or two other consanguinealy related nuclear families, usually siblings, with whom they share some economic functions. Elderly parents are also frequently coresident with their children, and indeed it is the rare household which is not three-generational. The next unit of inclusion in Ayta social organization is the camp or hamlet, which may consist of only a single household or of several. As noted previously, these hamlets are widely dispersed. An important feature of the camp or hamlet is the sharing of single hearth and the co-exploitation of respective members' swiddens, as well as cooperation in many other economic activities. This does tend to be somewhat variable, and is more the case in smaller such units than in larger ones. Nevertheless, as Fox (1953:247) noted, Ayta hamlet life is communal to an unusual degree. This stands in contrast to other Negrito groups such as the Agta; though Agta emphasize sharing, each Agta nuclear family maintains its own hearth (Estioko-Griffin and Griffin 1981). This is not at all the case among Ayta. Unless by themselves, or in the larger, more permanent settlements, nuclear families almost never cook and eat by themselves. Rather, cooking is a collective endeavor. The same applies to the procurement of food. Many swiddens within the research area are cleared and cultivated by two or three families. Alternately, families in a camp may procure food from each other's swiddens. A hypothetical three-family localized settlement group may have between four and seven swiddens under cultivation at once. In procuring food, members of this group may visit any of these swiddens on successive days. This is in part a function of task groups: there is a decided preference for working together in small groups of two or more. In few cases would a family be consuming only the products of their own swidden for any length of time. The Ayta preference for close kin to live together allows them to maintain this highly communal mode of operation. I suggest that the camp or hamlet, though not the most fundamental unit, is the primary operative unit in Ayta society. The formation and dispersion of camps or hamlets is in large part a function of the developmental cycle (*sensu* Goody 1958).

The third level of integration is the *sakop* or district. *Sakop* are recognized, named areas, averaging between 15 and 20 square kilometers in size. They are not discrete bounded territories. The inhabitants of a *sakop* tend to be closely related and a particular *sakop* generally contains a large proportion of an individual's kindred. Schlegel (1979:5) speaking of a similar, though discrete, territorial unit among the Tiruray, notes that:

> A neighborhood consists of a number of families, probably but not necessarily living in several settlements, which regularly assist each other in farming activities and rituals, and thus determine a subsistence cooperation group.

This is certainly one of the most important aspects of *sakop* membership. During large-scale planting or weeding operations (*mang'awis*) a large percentage of *sakop* members may participate.

Within the research area there were two *sakop*, Sawang/Maagə'agə' and Kalantas/Lomboy, with members of two other *sakop*, Malan'ak and Ambangan, being frequent visitors and having a high proportion of relatives in this area. *Sakop* Sawang included all of Maagə'agə', Bayabas, and several minor side valleys. *Sakop* Kalantas included Kalantas, Banu'a and Lomboy. People in Lomboy would call it *Sakop* Lomboy and include the inhabitants of Kalantas. The upland area between these two *sakop* was more neutral, the inhabitants of this area identifying with one or both. The relationship between these two *sakop* was tense during the period of field research due to a killing which had occurred some years previously. The other two *sakop*, Ambangan and Malan'ak enjoyed a particularly close relationship to Sawang, with kinsmen making frequent visits into these other areas.

Ethnographic accounts of the Ayta have generally attempted to force them into the nomadic hunting and gathering mold characteristic of other Philippine Negrito societies. This is an inaccurate characterization. Family groups do certainly move around frequently but, except in unusual circumstances, generally within the *sakop*. Movement between *sakop* may occur, particularly in cases of *sakop*-exogamy, where a family may shift residence from one *sakop* to another. Men in particular make frequent brief trips into other *sakop* in order to visit relatives.

There is a strong socio-political dimension to *sakop* residence, and each *sakop* has a *kapitan*. The *kapitan* is an influential individual who functions as a mediator of both internal and external conflict or who, in the case of open conflict, may assume a leadership role based on popular support. Obviously the term *kapitan* has been borrowed from the Spanish, and the assumption has been that the institution itself was imposed on the Ayta during the Spanish colonial period. Such an explanation is inadequate. The Spanish had notoriously bad luck in pacifying and settling Negritos throughout the Philippines until the very end of the nineteenth century. Their contact with Ayta in interior Zambales was limited to infrequent punitive actions. Administratively, they never did penetrate this area. Yet by the turn of the century this institution was firmly entrenched among the Ayta (Norbeck 1956, Reed 1904, Parker 1964). It is more likely that it has its origins in pre-Hispanic Sambal society, and is most directly concerned with conflict and alliance. The Ayta continue to espouse a strong ideology of conflict, though almost always stated in the idiom of defense. The *kapitan* is frequently mentioned in this context.

Finally, it is worth noting that among Ayta there is a rather rudimentary and highly variable development of what look like descent groups or lineages. Though not all Ayta within the research area adhere to this, there are others who vigorously espouse their relationship to a specific descent group. Within the three *sakop* of Sawang, Malan'ak and Ambangan nearly all individuals consider themselves descended from a single apical ancestor. In Kalantas and Lomboy the principle of descent seems considerably less important, though certain individuals and families do stress it. Ayta base this on a principle of asymmetrical cognatic descent, originating with a single male ancestor. By asymmetrical is meant that the descent of male and female offspring is reckoned differently: there is a strong emphasis on males tracing descent from a male ancestor, though there may be one (but never more than one) female link. A number of men are able to trace descent to the +4 generation. The line of descent for women is shallower and more variable, usually going back only to the +2 generation. The result of this difference between men and women in the reckoning of descent is that there is a concentration of men into fewer descent groups, while the descent group affiliation of women is considerably more diverse. Men always take on the surname of their father or maternal grandfather, while women invariably take on the surname of their mother, which may be derived from either the maternal grandmother or grandfather. By no means are genealogies known by all members of a descent group. It is generally older men who are the custodians of genealogical knowledge. Significantly perhaps, all descent groups appear to be of similar time depth, many of the surnames deriving from members of a single, specific kindred.

CHAPTER 3

PHYSIOGRAPHY

The Cabusilan Mountains of southern Zambales are physiographically complex and it is apparent that a number of orogenetic factors have shaped the present landscape. This has significance in that such factors, which determine slope, soils and so on, influence the nature of succession as well as limit the potential for exploitation of this environment by humans.

The Zambales mountain chain, of which the Cabusilan Mountains are a part, is geologically quite young, particularly the area from Mt. Pinatubo southward. The northern and southern halves of the mountain chain are of separate origin, though both are derived from the same orogenic mobile belt (Gervasio 1966:130-44). The northern half of the Zambales chain is formed from Cretaceous-Paleocene ultramafic plutonic material (Ibid.:131-33). The Cabusilan area is geologically much younger, its formation the result of Pliocene-Quaternary volcanic activity (Bureau of Mines 1963). The entire landscape is thus formed of igneous materials. Mt. Pinatubo, which in profile is clearly an extinct volcano, is composed of pyroxine andesite (Ibid.). The mountain area south of this, with the possible exception of the peaks of Mt. Negron and Mt. Cuadrado, is composed of pyroclastics, primarily a conglomerate in a tuffaceous matrix (Ibid.). A granite-like metamorphic basement complex is evident in some areas, primarily in riverbeds at lower elevations.

The topography of the research area is exceedingly rough, due in large part to the youth of the landscape. The mountains rise abruptly from the sparsely populated plains inland from San Marcelino. This area, covering approximately 68 sq. kilometers (26 sq. miles), ranges in elevation from under 200 meters on the western edge, to over 1700 meters on the summit of Mt. Pinatubo. The western watershed is bounded by three large mountains running north to south: Mt. Pinatubo (1,745 m/5,730 ft.), Mt. Negron (1,580 m/5,185 ft.) and Mt. Cuadrado (1,324 m/4,345 ft.).[5] Immediately to the west of Mt. Negron and Mt. Cuadrado is a large, deeply dissected upland area, ranging from 750 to 1050 meters. Cutting through the middle of this upland is a large precipitous valley: Maagə'agə'. Running northwest from the base of Mt. Cuadrado, intersecting Maagə'agə', is another large, nearly unpopulated area: the Kaopan ravine. It remains heavily forested. Much of the research area is characterized by numerous rocky escarpments and outcrops. The northwest edge of the area is bounded by the Lomboy "Plain," an extremely level area of fluvial origin composed of sand and rock almost completely devoid of trees. This area, which was visited by a contingent of U.S. Marines in April 1907, was described as "strewn with immense boulders, some of them many tons in weight" (Smith 1909:21). It remains this way today.

The composition of soils in the area is uncertain, though Oxisols, Inseptisols, Ultisols and Alfisols are probably all present. The soils on exposed, grass-covered ridges at higher elevations appear to be reddish Oxisols. Erosional slumping is very common in such areas. For the Lomboy area and onto the San Marcelino plain the "soil" has been classified as "Angeles Sand" (Alicante et al. 1951:29). It is the product of alluvial deposition, and lacks a developed profile. "It is gravelly sand, loose, and poor in organic matter. . . . The soil has a poor water holding capacity but [is] rather high in capillary action" (Ibid.:29).

A significant feature of the research area is that, the terrain being so mountainous except for Lomboy, there are virtually no depositional areas. Erosion is the single most important geomorphological process shaping the landscape. Even the Lomboy plain, which is the result of alluvial deposition, has clearly been subject to severe erosion at its margins.

There are two major drainage basins in this part of the western Cabusilan watershed. To the north is the basin draining the western slopes of Mt. Negron, the south and southwest slopes of Mt. Pinatubo (part of the Pinatubo radial drainage pattern), and the Kalantas area. To the south are those streams draining Maagə'agə', the north and

---

5. The Ayta names for these three distinctive peaks are, respectively, Pinatubo, Abu and Lingai.

west slopes of Mt. Cuadrado and the Kaopan Ravine. The streams forming the latter catchment area are particularly embedded, forming deep, very steep canyons.

1. Research area seen from northwest: Lomboy Plain in middle foreground, Mt. Negron left background, Kalantas extreme right background.

2. Central upland area: Lomboy Plain upper left, Kalantas upper right (circular forest area), Maagə'agə' at bottom. Note "fern-leaf" pattern of forest vegetation.

3. Kaopan Ravine: lowland forest to montane forest transition.

4. Maagə'agə'. Note extent of grasslands, *amukao* in foreground, small forest patches and burned areas.

5. People returning from visit to neighboring hamlet, upper Maagə'agə'. Ridge between western and eastern Cabusilan watersheds visible in background.

6. Mt. Pinatubo, seen from Kalantas.

7. Lomboy Plain, from photo taken by Smith in 1907.

8. Lomboy Plain in 1980.

9. Boys in search of honey, walking across Kapisan.

10. Early morning fire—Lomboy.

11. Ayta man setting out to collect sweet potatoes.

12. Putting the finishing touches on a new bow—Kalantas.

13. Dry season *haong*—Maagə'agə'.

14. Dry season *haong*—Kalantas.

15. Bayabas house cluster. Note proximity to swiddens.

16. Family in front of hut—Maagə'agə'.

17. Bayabas men setting out to work in swiddens.

18. Dancing to *guitara* music—Kalantas.

CHAPTER 4

CLIMATE

It is sometimes assumed that tropical climates in areas such as the Philippines are hot and wet with little variation. In fact, however, there is a great degree of climatic diversity in the Philippines, both regionally and within localized areas. An understanding of climatic factors is critical to the interpretation of any dynamic ecosystem. Wernstedt and Spencer (1967:62) remark that:

> To a very considerable degree climate controls natural vegetation, and together climate and the resulting biotic environment have profound effects upon soil development through weathering, leaching, eluviation and humus accumulations. Similarly, although lacking an absolute effect, the selection of crops and cropping systems and the resultant crop yields reflect prevailing climatic conditions rather clearly.

Climate has been a primary determinant of vegetational dynamics in upland Zambales and, by extension, has exerted a profound effect on the present Ayta agroecosystem.

In investigating both natural ecosystems and agro-ecosystems, whether from the perspective of succession theory or otherwise, it is important that limiting factors be defined and the degree to which these fluctuate around the mean be expressed. This has been a major problem of ecological description, as Winterhalder (1980:137-38) and Nieuwolt (1966:180) have pointed out. May (1973:109) states that "real environments are uncertain, stochastic," and that "Parameters are fluctuating randomly about their mean values." Not only should random fluctuation be considered; extremes are of particular importance. Selective pressures are most seriously felt during such extremes and, though they may occur only infrequently, they are nevertheless of major significance. Climate is among the more critical limiting factors which should be considered.

Perhaps the most important climatic factor to be taken into account is precipitation. The water balance of any terrestrial ecosystem--a direct function of precipitation--has significance for that system's primary vegetation, successional seres, and for any crops which are grown there. The spatial and temporal patterning of precipitation should therefore be made explicit when statements about succession are made or where the concern is with agro-ecosystems.

Richards (1952:139) states that "For the vegetation, the distribution of rainfall through the year is usually of far more importance than the annual total." Zambales is characterized by extreme seasonality in rainfall. The year is divided into two seasons of roughly equal duration. The months from December to April are characterized by extreme dryness, while from May to November it is extremely wet (see Figure 1). Few areas in the Philippines exhibit this to such a considerable degree. Zambales Province has one of the highest total rainfalls in the Philippines (Coronas 1920:75), averaging 3937 mm (approximately 150") in Iba, on the coast some 35 miles to the northwest of the research area.[6] Of such high rainfall areas in the Philippines, Zambales is unquestionably the most seasonal. Figure 4, a synopsis of 44 Philippine weather stations, shows this clearly. The implication is that when rain does fall, it is often of a torrential nature.

---

6. This is the only station in Zambales Province for which relatively complete records are available. The years 1902, 1911-1939 and 1960-1972 are the only years for which complete monthly data are available, and are incorporated where appropriate. It may be methodologically tenuous to extrapolate from generalizations derived from Iba data. However, I feel that this is sufficient for expressing the degree of variability characteristic of the region over time. As Iba is in an orographically similar situation to the research area, i.e., bounded on the east by high mountain ridges and peaks, this is perhaps less risky. Data from the Benguet Consolidated mining operation, approximately 5 miles southwest of the research area, are available from January 1976 to June 1980.

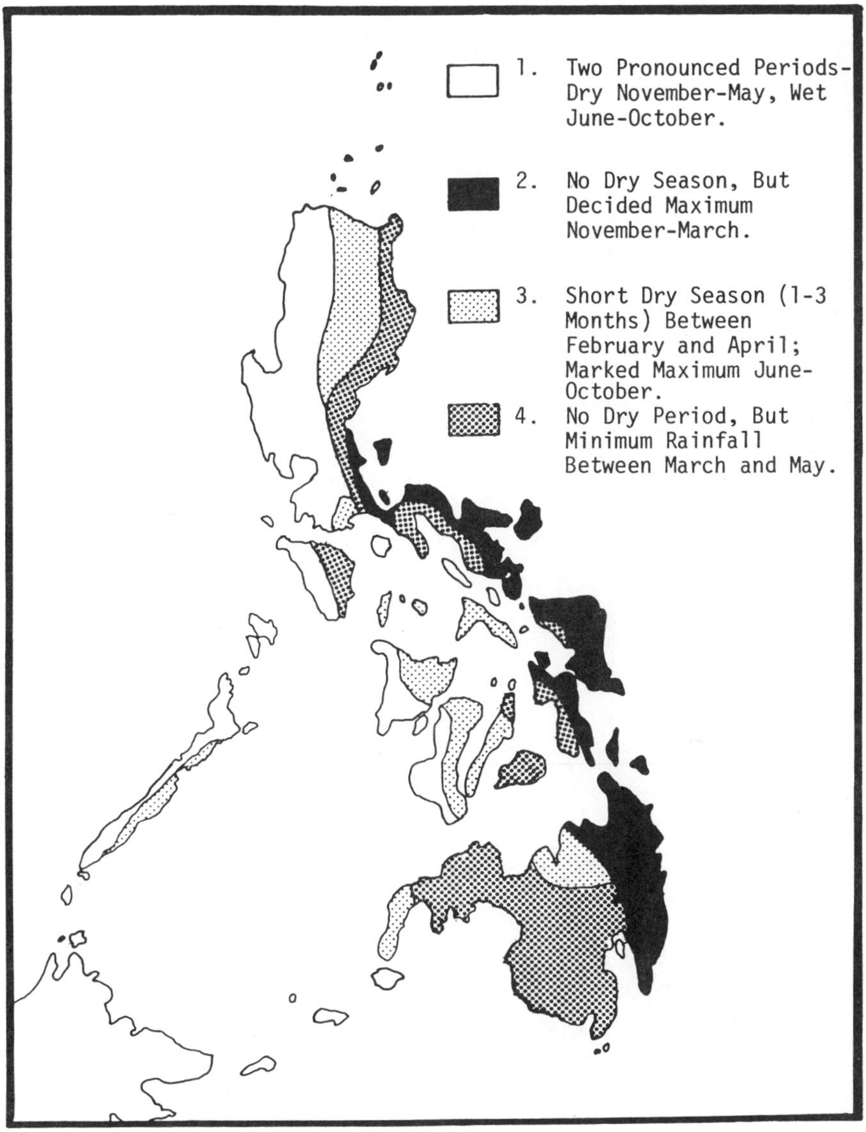

Map 3. Climates of the Philippines (Coronas 1920).

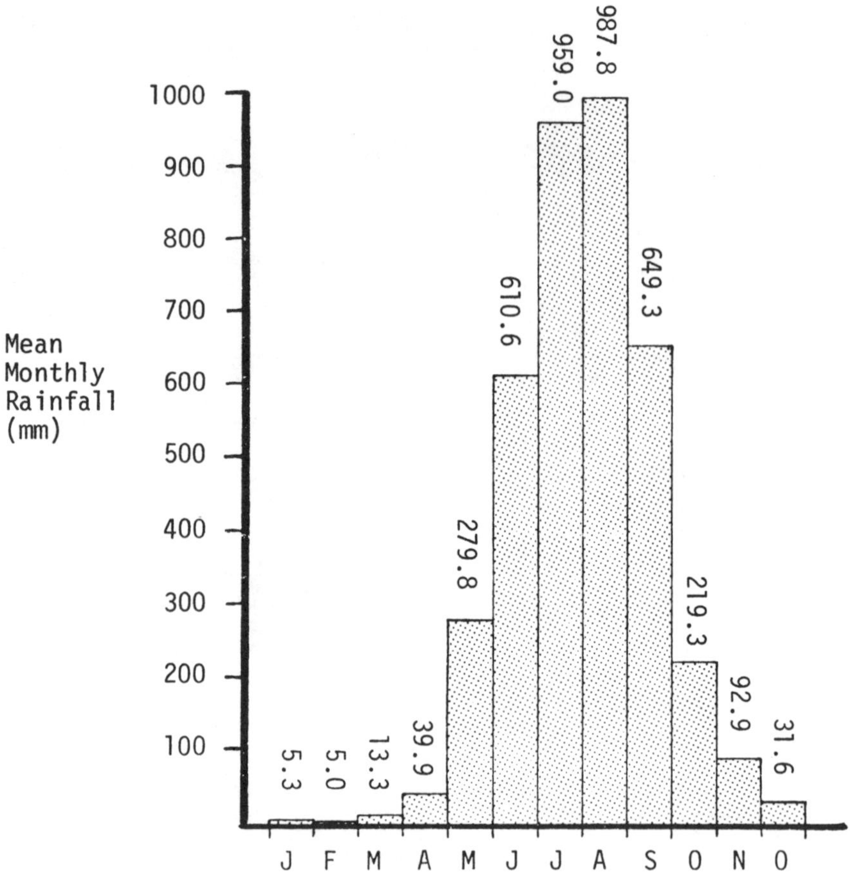

Figure 1. Mean Monthly Rainfall--Iba. Compiled from: PAGASA 1966-72, Philippine Commonwealth Weather Bureau 1903-38, Philippine Weather Bureau 1957-58, 1960-65.

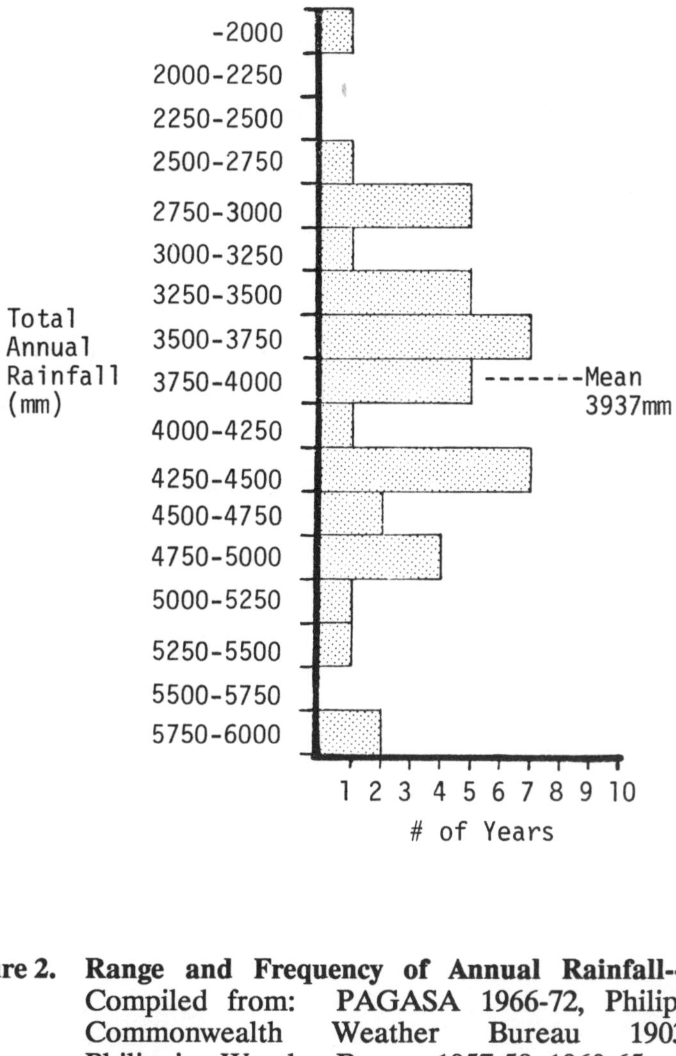

Figure 2. Range and Frequency of Annual Rainfall--Iba. Compiled from: PAGASA 1966-72, Philippine Commonwealth Weather Bureau 1903-38, Philippine Weather Bureau 1957-58, 1960-65.

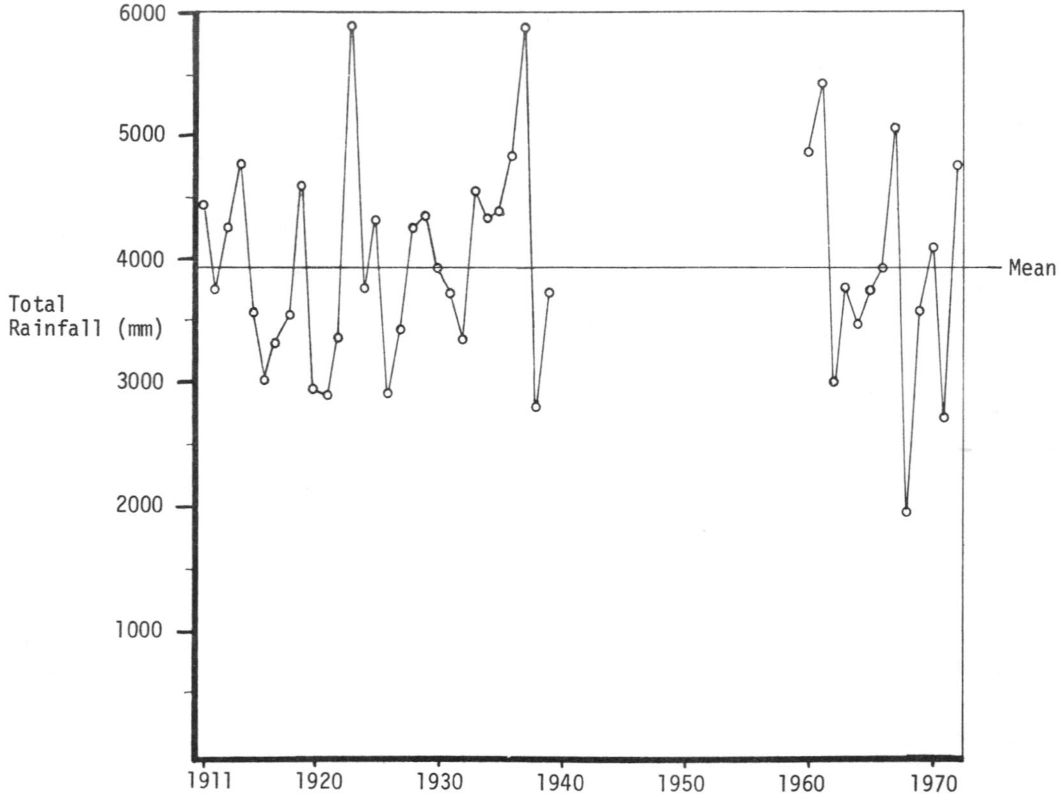

Figure 3. Iba Rainfall--1911-1939, 1962-1972. Compiled from: PAGASA 1966-72, Philippine Commonwealth Weather Bureau 1903-38, Philippine Weather Bureau 1957-58, 1960-65.

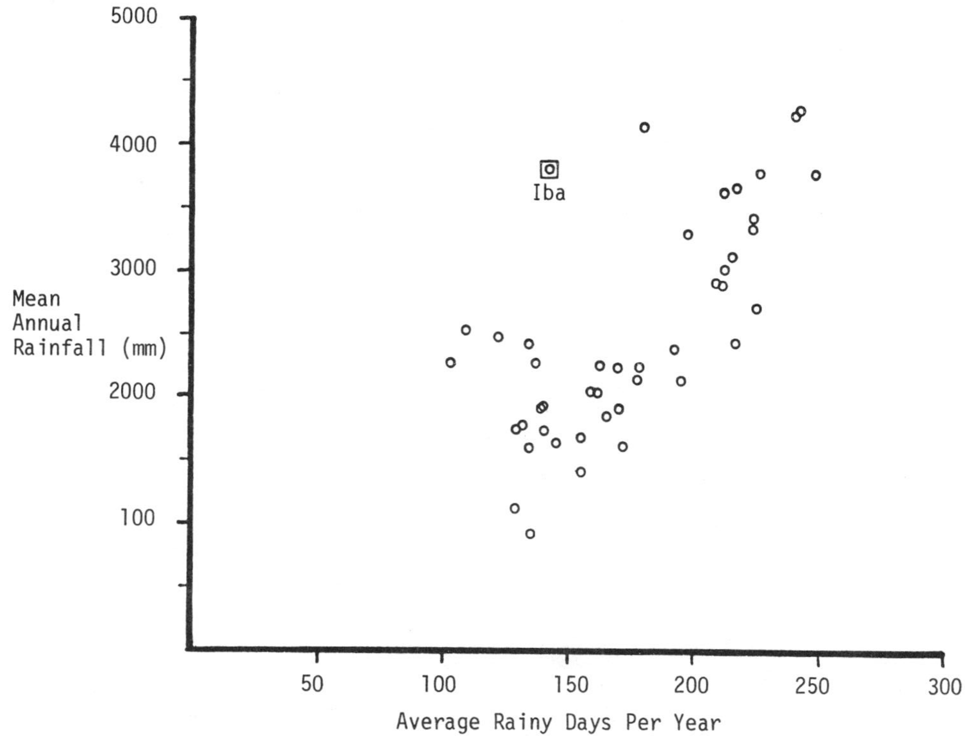

Figure 4. Rainfall: Mean Annual/Days per year--44 Philippine Stations.
Compiled from: Flores and Balagot 1969.

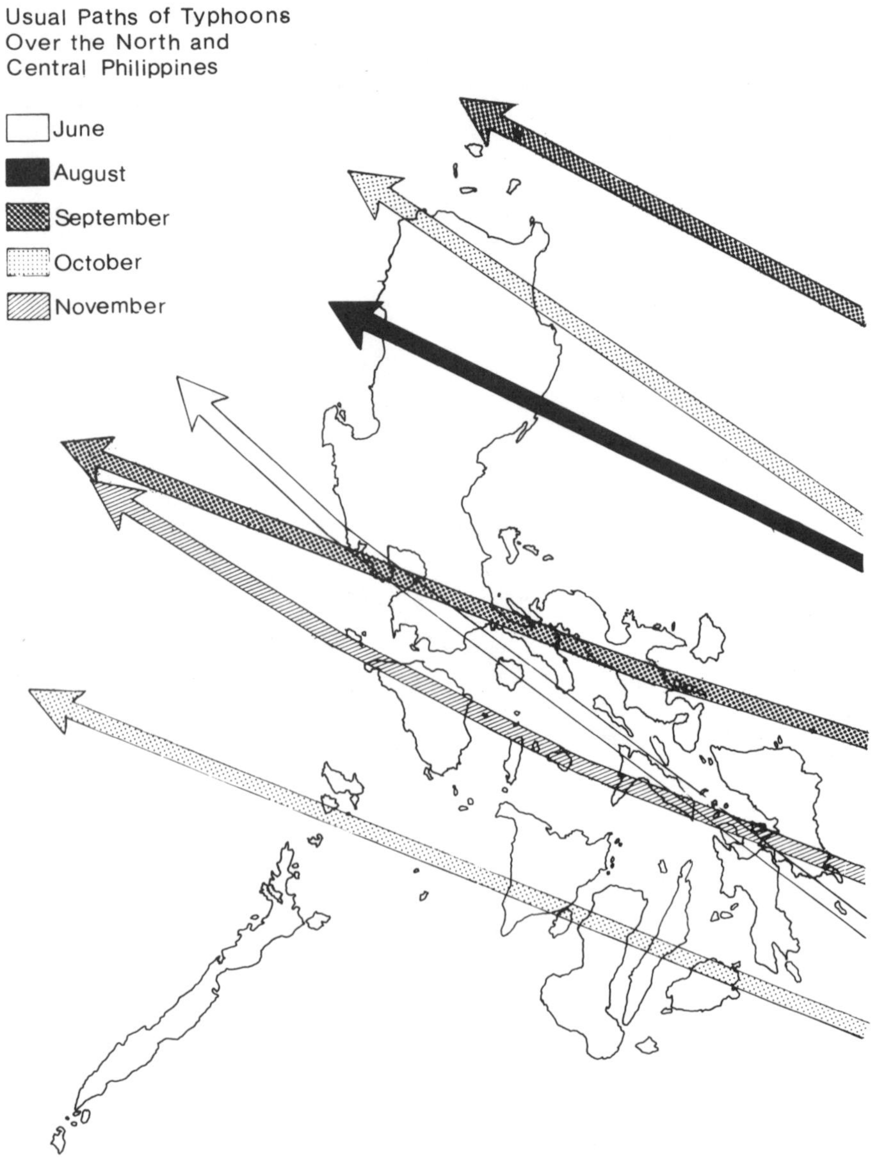

Map 4. Usual Paths of Typhoons Over the North and Central Philippines (Chief of Naval Operations, Aerology Section--1944).

Over 95 percent of the total rainfall in Zambales occurs between May and October. The dry months, on the other hand, are among the driest in the Philippines, averaging less than 16 mm of rain per month from January to April, with January and February averaging 5 mm. Months with no rain at all occur during this season with frequency. Data from the Benguet Consolidated mining operation indicate even more extreme seasonality, though longer term data would be required to confirm this. A four-year mean of 5309 mm (209") was recorded for the years 1976-1979. The dry season monthly means are equivalent to those recorded in Iba. Clearly, as Nieuwolt (166:177) points out, "Precipitation figures alone cannot very well indicate the dryness of a place." This has major implications for both the composition of climax vegetation and for successional communities.

The source of this highly seasonal rainfall pattern can be found in three interacting factors: the southwest monsoon, the orographic effect and typhoons. Several minor factors also contribute.

From approximately June to October the southwest monsoon affects western areas of the Philippines, particularly where high mountain ranges form a barrier. This monsoon has been described as a "maritime equatorial air mass" (Flores and Balagot 1969:164) which has its origin in the Indian Ocean region. It has a high moisture content, is very warm (Flores and Balagot 1969:165) and, according to Wernstedt and Spencer (1967:46), is "extremely persistent."

Exacerbating the effect of the southwest monsoon is the orographic cooling which results when this air mass encounters the Zambales mountains. Huke (1963:32) contrasts Iba with Tarlac, which is on the eastern side of the Zambales Mountains. Being in a rain shadow, Tarlac receives a yearly average rainfall of only 76 inches (1930 mm), while Iba receives twice this. Sandoval and Mamaril as well (1970:18) note that "the eastern foothills of the Zambales mountain range register the least amount of rainfall (in central Luzon) due to the shielding effect of the mountain range." Wernstedt and Spencer (1967:52) point out that:

> The amount of rainfall produced by forced orographic ascent will vary considerably and will depend on the height of the barrier, the temperature and water content of the air, and the rate of ascent.

Another factor is the angle of the barrier to the movement of the air mass. Huke (1963:31) notes that orographic cooling "is most pronounced where a mountain range lies directly athwart a prevailing wind blowing from a warm water body onto the land." Areas such as the island of Palawan are parallel to this air mass and are thus not greatly affected (Huke 1963:39). The Zambales mountains are almost perpendicular to the prevailing southwest monsoon.

The third major factor determining the seasonal distribution of rainfall in western Luzon is typhoons. Zambales Province is in that area of the Philippines which receives from 11 to 20 percent of the typhoons which pass through the country (Flores and Balagot 1969:73). The major typhoon belt runs through northern Luzon, a further one to three degrees latitude northward. However, even these northward tracking typhoons bring considerable rain to Zambales. Typhoons last from one to seven days (Kintanar 1973:6) depending upon their speed, proximity and trajectory characteristics. Those typhoons which do strike the Zambales area have spent their winds in passing over Luzon and are generally much decreased in strength. Only rarely might winds in excess of 100 knots be expected. However, they continue to dump torrential rains unabated. The height of typhoon season, from June to November (though they may strike any time of the year) is coincident with the southwest monsoon. It is the overlap of these two climatic features each year which, exacerbated by orographic lift, most contributes to the extreme seasonality of the Zambales area.

Other sources may also contribute slightly to precipitation. Throughout the year thunderstorms may occur as a result of convectional activity, though they generally increase in frequency during May, heralding the start of the rainy season. As Kintanar (1973:7) points out, these are generally "small-scale and short period disturbances." Thunderstorms may drop large amounts of rain in a very short time during their passage. They are unpredictable and, over a short period, may result in significant differences in rainfall amounts over a small area. Compared to other sources of precipitation in Zambales, thunderstorms are relatively less significant and seldom do anything to relieve yearly drought, as evidenced by the many months on record during which no rainfall is recorded.

Rainfall is the most significant limiting factor for the biotic environment in Zambales, at both ends of the spectrum. It is also the most variable. The marked pattern of seasonality in rainfall in the Zambales area has been noted. This being the norm, both natural vegetation and crops are adapted to such a regime. However, this pattern of rainfall may deviate from the norm in several respects. These deviations are amplifications of the basic pattern of seasonality rather than departures.

From year to year the quantity of rain which can be expected to fall in the Zambales area will vary. Using Iba data as an example, it was found that a variability of 16.9 percent around the annual mean rainfall characterized the area. This variability was biased toward fewer heavy years and more deficit years.[7] Years of exceptional rainfall probably coincide with years of heavy typhoon activity. Extremely dry rainy seasons do not occur. The rains always come; at least the records available so indicate.

It is during the annual dry season (*ka'ingitan*) that biotic elements face the most critical stress. The period from December to April is generally very dry, and consecutive rainless months are commonplace. The wettest January and February on record had rainfalls of only 35.3 mm and 28.5 mm respectively. The result is a drastically altered water balance from that which exists during the rainy season. Such a situation selects against plants lacking tropophytic[8] qualities and has a major effect on the regenerative ability of areas under fallow. The effects of this lack of rainfall are exacerbated by a tremendous increase in insolation, resulting in high rates of evaporation. IRRI (1974:4) estimates that monthly losses from evaporation are on the average equal to 100 mm of rain, while percolation through the soil is equivalent to another 100 mm. Dry season rates of evaporation is Zambales almost certainly exceed 100 mm. The actual amount depends situationally on soils, vegetation cover, aspect and so on.

---

7. 21.5 percent negative mean, 37.3 percent positive mean for the months June to September. Overall June-September mean variability is 29.9 percent. Only 27.3 percent of the time does rainfall fall within 10 percent of the mean.

8. A tropophyte is "A plant . . . which possesses various adaptations to enable it to survive a period of seasonal adversity (cold or drought)" (Monkhouse and Small 1978:303).

Zambales province does experience an extreme annual water deficit during the dry season, though this does not constitute a particular stress on upland agroecosystems. The Ayta cropping system has evolved in adjustment to this seasonal pattern. The reproductive strategies of native and endemic plants are also geared to this climatic regime. In fact, during this dry period rainfall might actually be stressful, at least with regard to agriculture. If such rainfall occurs during the critical swidden burning period, which is generally in April, the next year's cropping possibilities may be severely constrained.

Jackson (1977:498) has pointed out that "In areas with a marked seasonal regime, variability at the start and finish of the rains is particularly important." In Zambales these are certainly the periods during which the greatest variability is exhibited and when rainfall is least predictable. The months of May and October are most unpredictable. The rains may begin early--almost never late--and may end either early or late. May variability is particularly high at 68 percent. In only 6 percent of the years on record is this month within 10 percent of the mean of 279.9 mm. There are more years with a low May rainfall, the negative mean being 50 percent, with fewer heavy years, which have a positive mean of 106 percent. The range for the years on record is from 16.3 mm to 1073.4 mm. In May of 1976, 2505 mm was recorded at the Benguet Consolidated mining operation. With such wide variability, the timing of planting is somewhat flexible. If rainfall begins too early however--in April--some difficulty may be encountered from the resulting constraint on burning. Records from Iba indicate that this has occurred five times since 1910. October has an overall mean variability of 59 percent. The low recorded for this month was 15.2 mm, and the high was 1033.6 mm.

Temperature is another important climatic variable to consider, particularly in upland areas where considerable differences due to elevation may exist. Being within the tropics, mean temperatures in the Philippines are invariably high. However, significant variability may be "obscured by averages" (Nieuwolt 1966:171). Huke (1963:48) notes that areas in the Philippines with the climatic type characteristic of Zambales "are also characterized by the greatest seasonal temperature ranges experienced in the country."

Actual temperature data were not collected during research, so the seasonal range of temperatures which might be experienced within the research area cannot be stated with certainty. However it is possible to make a reasonable estimation by considering climatic data from the few upland stations in the Philippines (see Figure 5). These data were graphed along axes of elevation and temperature. Five types of temperature data were collected: (1) Mean temperature, (2) January mean minimum temperature, (3) lowest recorded temperature, (4) May mean maximum temperature,[9] and (5) highest recorded temperature. Only one source contained comparable statistics for upland areas (*Philippine Weather Bureau 1960-65*) and this covered a period of only six years. Some locations were represented by only two years of data; clearly inadequate. Three locations have considerably longer records: Iba (3 m), Baguio (1481 m) and Malaybalay, Bukidnon (642 m). Where supplementary data on extremes were available these superseded data from the years 1960-1965. The data for all these locations were then plotted. Five lines of regression, one for each class of statistic, were generated using the least squares method. The shaded portions on the outside of the minimum and maximum recorded temperature lines of regression represent the absolute extremes of temperature which might be expected. We are thus provided with an estimate of the bounds within which temperatures might be expected to fluctuate at different elevations within the research area. Locations such as Bayabas, at 940 m, may regularly fall below 14°C during colder periods of the year and at the extreme, possibly lower than 8°C.[10] Such temperatures may well be stressful to certain crops.

Several other factors should also be considered in constructing the climatic profile of a particular area. The presence of fog, mist and dew may be significant, particularly at higher elevations. Many Ayta in southern Zambales are living at and just below the ridgeline separating the Zambales and Pampanga watersheds. In such areas fog and mist are commonplace, particularly during the wet season. This may provide an important source of moisture for crops such as

---

9. Throughout the Philippines January is the coldest month and May generally the warmest. High temperatures may also be experienced during April. This effect is most marked in northern parts of the Philippine archipelago, in upland areas, and in areas less subject to maritime influences, such as inland valleys.
10. Brown (1919:215) recorded a reading of 9.5°C in 1913 at the 1050 m summit of Mt. Makiling, south of Manila.

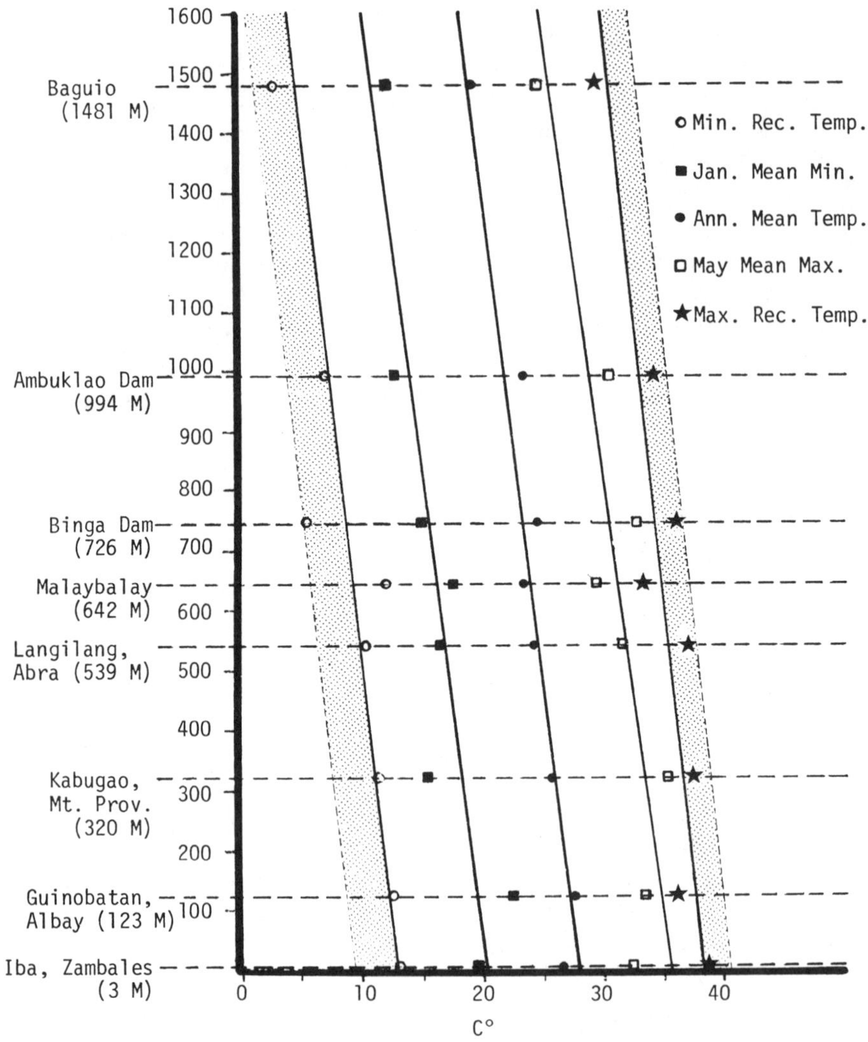

Figure 5. Elevation and Temperature: Least Squares Regression. Compiled from: Brown 1919, Coronas 1920, PAGASA 1966-72, Philippine Commonwealth Weather Bureau 1903-38, Philippine Weather Bureau 1957-58, 1960-65.

sweet potato. Wind is another important factor. At higher elevations in the Cabusilan mountains, wind is nearly constant. During the dry season such winds may contribute significantly to evaporation (Richards 1952:14). Typhoon winds may also be important. Such winds are usually much diminished by the time they reach Zambales. Every few years, however, winds gusting up to 80 or 90 knots can be expected; at least two such episodes have occurred since 1967. Much higher winds are likely to occur on rare occasions and their selective significance should not be underrated.

In any landscape, we would expect to observe a variety of microenvironments. Within the research area, elevation is the most significant factor in creating this mosaic. Temperature variation with elevation has just been demonstrated. Rainfall is another factor significantly affected by this variable. De Boer (1950), working in Java in the late 1940s, found a strong positive correlation between elevation and rainfall up to a certain altitude, after which it dropped off. Brown (1919) and Jackson (1977) further corroborate this. The drop-off effect is most pronounced on higher peaks, perhaps in excess of 1000 m, and is significant only on very high mountains.

As noted, it is departures from mean values which are of greatest selective significance for the biotic elements of any ecosystem. It is such stochastic fluctuations to which populations must gauge their adaptive strategies and during such episodes that selective pressures are most felt. The climatic regime characteristic of Zambales is significant in that, with wet and dry extremes, the limits of tolerance of biotic elements are being stretched at two extremes. Furthermore, with deforestation, the impact of such extremes may be exacerbated due to higher evaporation rates, increased run-off, increased percolation in certain types of soils, and more extreme temperatures. These may exert severe selective pressure on certain components of the biotic community and thereby favor highly tolerant disclimax species.

CHAPTER 5

VEGETATION

Due both to the variable nature of the terrain and to the radically altered nature of the contemporary environment, the research area supports a variety of vegetational communities, mostly of very mixed distribution. It is not the purpose of the present section to discuss the genesis of this diversity. For now the primary concern will be with the distribution, floristics and structural characteristics of the major vegetational associations.

The research area was presumably at one time entirely forested, with the possible exception of the Lomboy area: it is possible that this grass-covered plain represents an azonal edaphic climax. Two types of evidence indicate the extent of previous forest cover in southern Zambales.

There is first the historical record. A number of statements by early Spanish chroniclers indicate that the region was once heavily forested and was a rich source of timber. This was particularly true of the eastern watershed in Pampanga, an area which today is almost completely grass-covered. The Malong revolt of 1660 began as a strike by some 1,000 woodcutters working in the mountains of Pampanga and Bataan. Wood from the area was used extensively in the construction of ships and houses (Cortes 1974:15, 147; Larkin 1972:27). Accounts of Sambal[11] timber cutting go back very far. Zuniga stated in 1803 (Zuniga 1973:346) that:

---

11. The Sambal are the ethnolinguistic group traditionally inhabiting coastal and intermediate upland areas in Zambales Province. Today they are located almost exclusively in coastal areas. In the southern part of the province they have been displaced by Ilocano and Tagalog migrants, beginning in the early nineteenth century.

> The mountains . . . provide much costly lumber not only for building good houses with but also to furnish attractive items in their trade with Manila. All of these benefits furnished by the mountains provide them with some livelihood like cutting wood.

He also noted that the Sambal carried on lumber trade with Manila using log rafts for transport. Speaking of trade near Bagac on the west coast of Bataan, Perez (Blair and Robertson 1909, 47:292) wrote in 1680 that:

> They [the Sambal] live in their rancherias whence they get Molave wood (*Vitex parviflora*) in abundance. . . . they apply themselves more earnestly in cutting said timber than in farming their fields. They get considerable help for cutting said wood from the blacks of the mountain.

Several other writers note that in the early colonial period the Sambal paid their tributes to the Spanish with lumber.

The presence of remnant forest patches and isolated individual trees in the research area also indicates that forest cover was at one time more extensive in southern Zambales. Throughout the area, in small valleys and on protected slopes, patches of primary forest remain, surrounded by grasslands or other secondary communities. This can be clearly seen by reference to the vegetation map (Map 5; see also Plate 2): a "fernleaf" pattern of forest (*sensu* Paijmans 1976:93) is evident. Scattered individual trees are also present, emerging out of grassland and other secondary associations. Dipterocarps and other primary forest tree genera usually require deep shade in order to germinate and become established (Brown and Mathews 1914:337). Presumably then, areas where such trees are present were once forest-covered.

Together this evidence indicates that the research area was at one time heavily forested, and has subsequently undergone significant transformation on a major scale. Clearly a number of very dynamic vegetational processes have been occurring. Those processes will be considered in subsequent chapters. For now the concern will be with a definition of the present biotic environment, and a review of those factors which determine vegetational floristics and structure.

A primary determinant of the floristics and structure of vegetation in upland Zambales has been the extreme seasonality in rainfall. As noted in the previous chapter, it is the distribution rather than the amount of rainfall which is of greatest significance. Brown and Mathews (1914:418) state that the "relation between the distribution of forests and rainfall is due to the seasonal character and not to the total amount of rainfall." The pattern of rainfall in Zambales has a decided impact on species composition: a great number of species characteristic of other Philippine forests are absent there. Further, these forests tend to be less diverse than those elsewhere in the Philippines. Hubbell (1979:1300), speaking generally of tropical dry forests, states that "the dry forest has only a third to half the number of tree species characteristic of moister forests" (see also Flenley 1979:3). The phenology[12] of species present is also affected by the pattern of rainfall. Leigh (1975:73) notes that "Brown and Mathews (11, p. 521) found a seasonal rhythm to the growth of trees in Philippine rain forest, with sharp minima in the dry season, when water was short and in the cloudiest portion of the rainy season when evaporation was lowest." Finally, extreme seasonality engenders a deciduous element in the forests of Zambales. As Merrill (1945:82) observes, "Leaf fall is much more marked seasonally in those regions where long dry seasons prevail."

Forest composition and structure is also influenced by the physical character of the landscape. As noted, the deeply dissected upland environment in southern Zambales is characterized by numerous steep slopes and cliffs, canyons and talus areas. This engenders a high degree of micro-environmental variation. Huke (1963:92), speaking of the composition of Philippine forests, notes that they show "considerable variation depending upon local site factors." Speaking in reference to the climatically and edaphically similar forests of Bataan, some 35 miles south of the research area, Brown and Mathews (1914:433) note that "the rough topography has

12. By phenology is meant the morphological and physiological life history characteristics of particular plant species, primarily as these relate to external environmental factors such as seasonal changes in the availability of moisture, sunlight, and so on.

resulted in a very uneven development of the stand in different situations." They write (Ibid.:432) that "Due probably to the pronounced dry season and rough topography, the forests of Bataan are more complex in composition than are the forests of northern Negros."[13] This has resulted in the smaller size of the dipterocarps and a "larger number of smaller-sized trees of other families which enter the dominant story" (Ibid.:432). Further, they state that "The poorer development of the main story, resulting in the entrance of a larger number of species other than dipterocarps, has produced well-developed second and third stories" (Ibid.:435). Undergrowth also tends to be more dense.

Elevation is another important determinant of forest floristics and structure in Zambales, producing a clear pattern of vegetational zonation. Zonation on tropical mountains has been a focus of research for a very long time, and a large corpus of data has been amassed. A number of factors have been cited as controlling this. Temperature, moisture, ultraviolet radiation and edaphic conditions have been the most frequently cited,[14] but it is not presently clear how these factors interact or precisely how they determine zonal changes. Zonation, though it may be casually observable, must be viewed with some care. Flenley (1979:4) points out that "vegetation, if it obeys any laws at all, obeys statistical ones. Any zones, therefore, can only be regarded as areas of high probability for the occurrence of a particular vegetation type (e.g., Walker and Guppy 1976)."

Zonation in the tropics is most apparent with regard to forest structure. Several pantropically uniform zonation patterns have been recognized. Richards (1952:354) notes that "the chief change in the structure of the vegetation with increasing altitude is a decrease in height, accompanied by a reduction in the number of tree stories." Virtually all studies of zonation on tropical mountains corroborate this observation. Lowland forests tend to be three-storied, montane forests two-storied, and mossy forests one-storied.

Zonation also has a floristic component. Grubb (1977:83) points out that, "There is virtually always a marked fall in species

---

13. The forests of northern Negros are, or were before removal, some of the richest and most productive in the Philippines, particularly with respect to the presence of great numbers of large Dipterocarps.
14. See Flenley (1979:5).

richness with increase in altitude. . . ." Also characteristic of Southeast Asia is a "strong contrast in geographical affinities between the lowlands and the mountain flora" (Richards 1952:348). Lowland vegetation is Indo-Malaysian while, particularly in the Philippines, montane vegetation is characterized by numerous north temperate and Australian taxa.

Physiography may have some impact on the pattern of zonation. According to Merrill (1945:84), "The line of demarcation between the low-altitude high forests and the medium-altitude mid-mountain forest may be sharp or rather gradual, for much depends on topography, presence or absence of ravines, steepness of slopes, and other factors." Further, in Zambales the zonational pattern is depressed somewhat, the various zones occurring at lower elevations. This is known as the Massenerhebung effect, first documented in the Alps, wherein "zone boundaries tend to occur at lower altitudes on small isolated peaks near the sea than on larger mountain masses inland" (Flenley 1979:5).

Considering the factors of topography and zonation cited above, and given its history of continual deforestation, the Cabusilan area today exhibits a complex mosaic of various types of primary and secondary vegetational associations. These can now be defined. It should be noted that, particularly when referring to secondary vegetation, elements of the following typology may represent stages in a single successional sere: as Paijmans (1976:96) notes, "one runs the risk of describing 'associations' that are in fact only seral stages." For now the purpose is to present a synchronic description of the vegetation. Successional dynamics will be discussed in subsequent chapters. It should not be assumed that the following categories are clearly differentiated. They are primarily modal types around and among which a great deal of variability exists. It is also worth pointing out one critical difference between primary and secondary vegetation: the high percentage of endemism in primary forest, as opposed to the low percentage characteristic of secondary vegetational associations. The percentage of endemism in Philippine forests is a very high 84 percent, while for secondary areas it does not exceed 10 percent (Merrill 1926:69). As Dickerson (1928:145) notes:

> The percentage of endemism is very low in the settled areas at low altitudes, in the open grasslands, and in the

> secondary forests, a condition which indicates destruction of the original primary forests and replacement of endemic by introduced species.

Merrill (1926:68) corroborates this point, noting that secondary species spread "after man had prepared the habitat to which these open-country species are adapted. . . ."

Seven types of vegetational community can be recognized within the research area. Three represent primary vegetation, three are secondary and one is a somewhat anomalous mixed category These include:

1. Lowland forest
2. Montane forest
3. Mossy forest
4. Secondary forest
5. Grassland
6. *Amukao*
7. Riparian vegetation

Lowland forest

Undisturbed lowland primary forest is all but extinct within the research area, due primarily to heavy logging in previous years. Only scattered stands and isolated trees remain. The only areas in which significant portions of this forest remain are on the western slopes of Mt. Balitog[15] and in the lower portions of the Kaopan ravine.

Philippine lowland forests are, with few exceptions, dominated by members of the dipterocarp family. The extent to which these dominate is remarkable. Whitford (1906:708), investigating one stand in Bataan, found that 60 percent of the trees in that stand, and 68.8 percent of volume, were composed of six species of dipterocarp. Three species[16] were found to form 45.9 percent of the stand and 51.9 percent of the volume. In another plot (1906:637) a single species (*Dipterocarpus grandiflorus*) composed 15 percent of the stand. This

15. Called *Balitok* by the Ayta.
16. *Dipterocarpus grandiflorus*, *D. verniciflus*, *Pentacme contorta*.

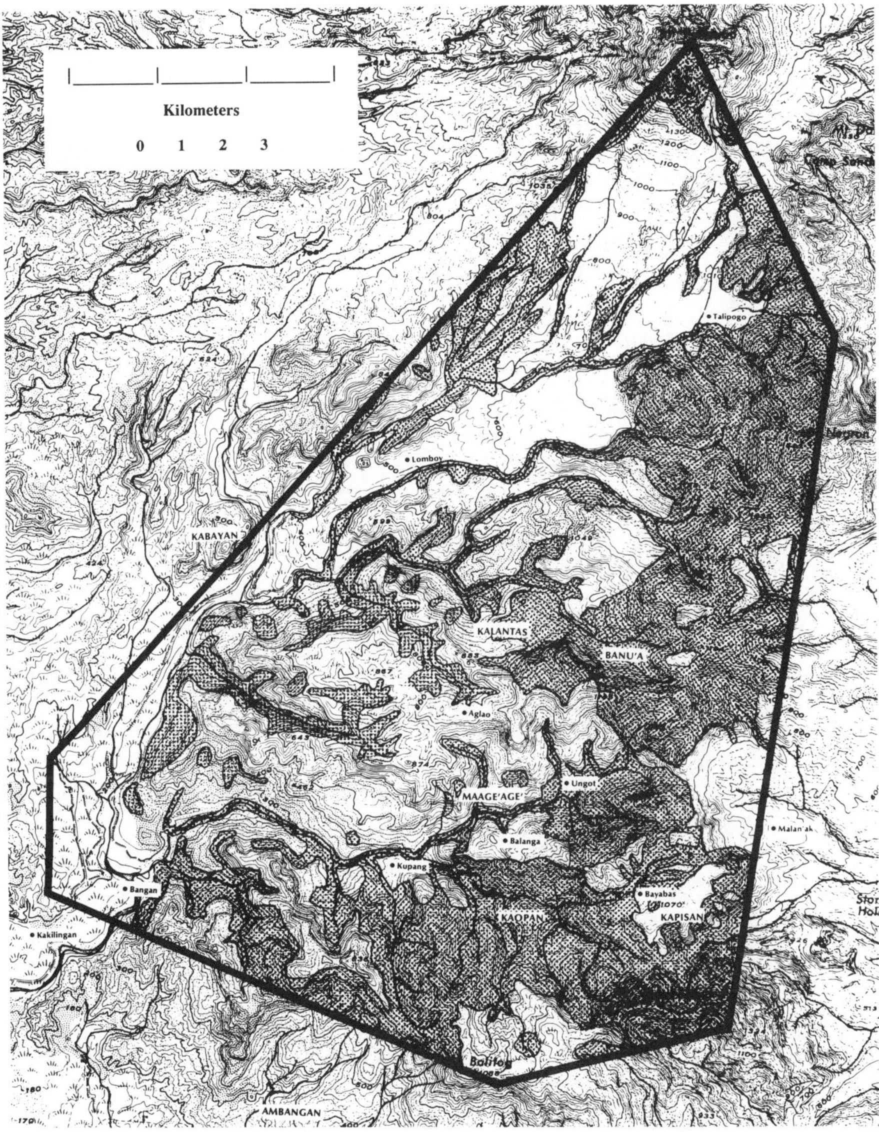

Map 5. Vegetation Distribution: Shaded area--primary/mature secondary forest vegetation; unshaded area--grassland/amukao/young secondary vegetation.

may in part be due to the effects of the extended annual dry season, which selects against many taxa which occur in climatically and edaphically wetter areas.

The lowland forest characteristic of the Zambales area is that classified by Whitford (1911), and others up to the present, as the apitong-lauan type.[17] According to Huke (1963:94), "The apitong-lauan subtype, or mountain lauan type, occupies the mountain slopes between the well-developed stands of lauans below and the mid-mountain forests above." It is found "where drier climatic or edaphic conditions prevail" (Ibid.:92).

It is likely that more detailed investigation would reveal several subtypes of lowland forest within the research area. In Bataan, Whitford (1906:414) found very different species composition on hill slopes and river terraces. On steep slopes of approximately 40° he found "fewer tall trees and many more saplings and seedlings" (Ibid.:413). Further, "In many places . . . the disintegrated portion of the rock is at or near the surface; in such places the vegetation is thin. Here deciduous or evergreen trees with xerophytic structures are common" (Ibid.:426).

Montane forest

Considerable stands of primary montane forest remain in the research area, particularly on the higher slopes of the larger mountains, in the Kaopan area, and on the ridge area bordering the eastern watershed. This type of formation generally becomes conspicuous at around 600 meters, though this transition is gradual and variable. As one ascends, large trees characteristic of lowland forest decrease gradually in frequency, density, and size. The upper limits of this formation are more variable. On the lower ridge areas bordering the eastern watershed the upper limits may be depressed to about 1000 meters, while on the higher mountains, it appears to go as high as 1400 meters. This upper transition zone also appears to be much more abrupt than the transition from lowland forest to montane

---

17. Apitong--*Dipterocarpus*, *Shorea*; Lauan--*Pentacme contorta*, *Shorea spp.*, *Parashorea*.

forest. Grubb (1977:84) notes that such a transition is closely associated with the cloud layer.

The montane forest characteristic of the research area is generally similar to the "tanguile-oak type"[18] defined by Whitford. He states (1911:25) that "The forests of this type cover the area extending from the upper limits of the lauan and lauan-apitong types to the lower limits of the mossy forest type in the higher portion of the mountains." The vegetation here is less tropophytic than that found lower, and the deciduous element is less significant.

Several trends are evident in the transition from lowland to montane forest. First, these forests tend to be "less diverse than lowland forest" (Flenley 1979:31). Floristically, this transition is marked by the gradual disappearance of the dipterocarps and the increasing dominance of mid-mountain taxa. Whitford (1906:642) found in a montane forest plot in Bataan that dipterocarps comprised only 16 percent of the stand for trees exceeding 4 meters in height. This decreases further with increasing elevation. He also reported (1911:25) that "Some species that usually are found along streams in the types of lower altitudes occur in deeper sorts of the tanguile-oak type away from the streams."

One very conspicuous feature of montane forests is the increasing abundance of ferns, epiphytes and other herbaceous vegetation with increasing elevation (Merrill 1945:84; Brown 1919:77). The tree ferns, *Cyathea spp.* are much more conspicuous, particularly in valleys and along streams. The rattans of the genera *Calamus* and *Daemonorops* are also common, though mostly as young specimens in the field layer; mature specimens have been greatly depleted. The fern *Asplenium nidus* and the climbers *Pothos hermaphrodites* and *Freycinetia sp.* are abundant.

Montane forest differs from lowland forest not only floristically, but structurally as well. Merrill (1945:83) writes: "These mid-mountain forests differ from low-altitude ones, aside from differences in individual species, in that they are essentially of a two-story type. . . ." The canopy tends to be lower, about 25-30 meters (Sci. Ed. Center 1975:32) as opposed to 40-50 meters in lowland forest.

18. Tanguile--*Shorea polysperma*; Oak--*Quercus spp.*

Mossy forest

Mossy forest continues to thrive nearly undisturbed on the higher slopes, ridges and summits within the research area. The elevation at which it begins is variable, from slightly lower than 1000 meters, as on the ridge east of Bayabas, to as high as 1400 meters on the upper slopes of Mt. Negron. The lower boundary of mossy forest tends to be very abrupt and as pointed out by Grubb (1977:84) and Brown (1919) this type of forest is "confined to the cloud belt" (Brown 1919:103).

Mossy forest is strikingly different from the types of forest found at lower elevations. It is characterized by growth which is stunted and extremely tangled, due to wind and high humidity. Epiphytic plants, ferns, mosses and liverworts grow in abundance. Merrill (1945:85) describes the "thick blanket of moss . . . saturated with moisture," and the "great profusion of mosses, scale mosses, lichens, ferns, orchids, and other small flowering plants." Structurally mossy forest is single-layered, the very even canopy generally not exceeding 6-8 meters. Straight-boled trees are absent. As one ascends and nears the peaks and ridges the canopy lowers even more until, in some places, a shrubby "elfin woodland" is encountered.

Another significant feature of mossy forests is the xeromorphic character of many of the species present. Leaves are thick, leathery and short-lived. This is a very "expensive" strategy in terms of energy and nutrients, and diverts production from the development of woody parts (Grubb 1977:87, 102).

A number of conditions account for this distinct forest type, though there is presently some disagreement as to the relative significance of particular factors. First, these forests are generally found in the steepest and most precipitous areas. Such areas are exposed both to extreme climatic conditions and to slides. Perhaps the prime controlling factor is that of moisture. As Brown (1919:98) points out, in these areas "relative humidity is high and evaporation low." Soil moisture content is also extremely high. As a result, podzolization has been the primary soil building process. The resulting podzolic soils are highly acidic and of very low fertility. At higher elevations temperatures are also low, relative to lower elevation areas. Other factors are high winds which, particularly at certain times of the year, may be nearly constant, and decreasing illumination and insolation due to frequent cloud cover.

As pointed out by Flenley (1979:3) species diversity decreases significantly with elevation. Relative to montane forest, the diversity of species in mossy forest is lower still. With regard to floristic composition, there appears to be a striking degree of generic uniformity in Philippine mossy forest, particularly within Luzon. Also, there are strong north temperate and Australian taxonomic affinities in mossy forest species (Dickerson 1928:145-146). Tree ferns, *Cyathea spp.*, are common, particularly near water courses, as is a ground dwelling Pandanus, and lone wild bananas. According to Whitford (1906:664) members of the families Melastomaceae and Rubiaceae are common components of the herb/shrub vegetation. *Rhodedendron sp.* and *Dalbergia sp.* are also common. The vine *Freycinetia sp.* is very conspicuous as is the scrambling bamboo *Schizostachyum sp*. Pitcher plants and numerous genera of orchids are also present. Ferns are particularly common. Finally there are the mosses, which cover virtually every branch. Successional vegetation at these high elevations is mostly dominated by *Homolanthus sp.*, *Miscanthus sp.*, *Rubus sp.* and *Histopteris sp*.

Secondary forest

Of all the vegetative associations so far discussed, this is by far the most variable and difficult to succinctly define. Secondary forest is extremely variable both floristically and structurally. It is a very broad category, ranging from mixed communities of secondary trees, grasses and wild bananas to mature communities superficially indistinguishable from primary forest. Such areas have a wide and very diffuse distribution, though they mostly occur in valleys and on protected slopes. All such secondary forest associations, called *kal'anan*,[19] are the result of human activity, i.e., the removal of primary forest vegetation.

Secondary forest is distinguishable from primary forest both structurally and floristically. With regard to structure, large, straight-boled trees such as the dipterocarps are completely absent except as

---

19. *Calaanan* in Tagalog--see Whitford (1911:14).

scattered emergents. Trees in secondary forest are short, usually about 5-8 meters, and seldom exceed 10 meters. Merrill (1945:60) notes that "Many of the trees have short trunks and these often branch freely like shrubs. . . ." Further characteristic of secondary forest associations is that trees are much more densely packed than those in primary forest. These trees are also more short-lived: Richards (1952:385) notes that "The small maximum height and diameter reached by secondary forest trees is in no way connected with their rate of growth, but is a consequence of their short life." Finally, as Brown and Mathews (1914:452) note, "The second-growth trees are very intolerant of shade and form only a very light canopy." The dense shading, characteristic of primary forest, does not occur, allowing the existence of a much denser ground layer of grasses, herbs and shrubs.

In terms of general floristic features, the most significant is the low percentage of endemism in secondary forests. Merrill (1945:63-64) states that "In a recently established secondary forest endemism may be nil, but as these forests attain age the percentage will gradually increase, although nowhere will it be very high--certainly not over ten percent." Richards (1952:386) remarks that many species "are pantropical or range throughout the whole neotropical or paleotropical region; very few have an endemic or localized distribution." Another general feature of secondary vegetation is the very low diversity, particularly at the generic level. Merrill (1945:60) states that "one frequently finds that the great mass of the rapidly growing shrubs and small trees belong in relatively few genera." In the Cabusilan area between one-third and one-half of the trees in any given secondary association are members of the genus *Ficus*.

The reproductive strategies and phenology of secondary tree species are also strikingly different from those of primary forest species. Secondary species are generally prodigious seed producers. Janzen (1979:40), speaking of *Ficus*, notes that they "probably generate the most thorough and extensive seed shadows found in any vertebrate-dispersed tropical perennial." Growth is also extremely rapid. Whitford (1911:14) remarks that early secondary tree species "are capable of producing seeds within a year or two after germination. Brown and Mathews (1914:454) state that "The seeds are usually small and are readily dispersed by birds or by the wind, and nearly all of the second growth species grow very rapidly."

Secondary tree species are, as noted, light-loving and intolerant of shade. They grow and reproduce rapidly before they can be shaded out. Merrill (1945:64) notes that when an area has been cleared:

> the habitats were provided wherein those types of shrubs and small trees that make up the secondary forests not only established themselves but could reproduce themselves and thrive without being overshadowed and eliminated. . . .

With regard to floristic composition, a variety of small soft-wooded species are common. As noted, the genus *Ficus* is heavily represented in all secondary areas. The genera *Oreocnide*, *Melanolepis*, *Antidesma*, *Lagerstroemia*, *Macaranga*, *Albizzia*, *Parkia*, *Premna*, and *Leuceana* are also common. In the field layer, various grasses are present. Most common, particularly at higher elevations, is *Miscanthus*. *Imperata* is also abundant. The wild banana, *Musa sp.*, is extremely common, either as single plants or in clumps. The combination of softwood trees, *Miscanthus* and *Musa sp.*, is found very frequently in second growth areas. Also extremely common in secondary forest is *Chromolaena odorata* (*Eupatorium*). *Aralia bipinata* and the scrambling bamboo *Dinocloa sp.* are also often seen. The Zingiberaceae are well represented. At lower elevations *Leuceana glauca* and *Stachylarpheta sp.* are common. Finally, numerous species of wild yams (*Dioscorea spp.*) are found in secondary forest areas.

Grassland

One of the most striking features of the research area are the large expanses of grassland found there. Approximately 35-40 percent of the total area is covered by grassland associations, several types of which exist. Such communities occupy a wide variety of physiographic positions; level lower elevation areas, intermediate valleys and high mountain slopes and ridges. The occurrence of these grasslands is almost entirely the result of human activity, though in some cases edaphic factors may be important. As Huke notes (1963:101), "Man has been the principle instrument in the creation, propagation and maintenance of the Philippine grasslands. . . ." Richards (1952:335) states that "There is no evidence that any of the grasslands of southeastern Asia is a climatically determined climax community." Whyte (1974) maintains that there were no grasslands in Southeast Asia before the advent of widespread agricultural clearing.

Once established, these grasslands are subjected to repeated annual burning, perpetuating their existence and increasing them at the margins. Large expanses may be burned in a single event and such events occur repeatedly during the annual dry season. The result is that, within the Cabusilan area, a significant percentage of the total area is burned annually. Approximately 18 percent of the total research area was burned during the 1980 dry season. As such these grasslands become "deflected successions" and "stable biotic climaxes" (Richards 1952:337).

The impact of extreme seasonality in Zambales is most clearly seen with respect to grassland. During the long annual drought these areas become desiccated and tinder-dry. Fires start easily, spread quickly, and cover large areas. As the dominant grasses possess dense underground rhizomes, they are not substantially affected by burning. However, herbs, shrubs and tree seedlings generally do not survive. As Walter (1973:68) notes, "Grass and woody species are antagonistic plant types," due to (1) root systems, and (2) water economy. It is in this manner that the pyrophytic grasses achieve stable dominance.

Other features of grasses allow their wide dissemination and persistence. Like so many other secondary species, they are prodigious seed producers. These seeds, borne on light tufts, are dispersed far by the wind. The two dominant species *Imperata* and *Miscanthus* also reproduce "by extension of a very vigorous rhizome system" (Holm et al. 1977:64). Further disturbance only aids reproduction. "Slashing, burning, defoliation, grazing and the addition of nitrogen stimulate flowering" (Ibid.:64).

Imperata is the most notorious of pantropical grasses. It has been observed that *Imperata* "can withstand long dry spells on light soils and will tolerate waterlogging on heavy soils. . . . *I. cylindrica* can occupy any soil type where sufficient moisture exists to support growth" (Ibid.:64). It invades early in succession and forms a dense sod (Huke 1963:99,103). It is highly drought resistant (Ibid.:99) and its high nitrogen requirement further gives it a competitive edge (Eussen et al. 1976:4). Finally, "The underground rhizomes of the grass have sharp hard points which actually enter . . . roots . . . and may grow within them for distances up to 60 cm" (Holm et al. 1977:69). However, it is a "light-loving plant which can be 'shaded out' under a heavy canopy" (Ibid.:66).

Miscanthus is equally abundant within the research area, particularly at higher elevations. It is one of the few plants which can compete successfully with *Imperata*, which it does at several stages in its life history. Its reproductive strategy is essentially the same as that of *Imperata*: seasonal dispersal of large numbers of windborne seeds, and the growth of a dense network of underground rhizomes. Once established, *Miscanthus* effectively shades out *Imperata*, being two to three times greater in height and occurring in dense clumped stands. Were *Miscanthus* not present it is likely that *Imperata* would colonize those areas. Though *Miscanthus* is light-loving (Yamada 1975:397) it can exist and reproduce in low light conditions as well. Yen (1976:101) reports *Miscanthus* growing along streams in the densely forested Tasaday reserve area in southern Mindanao.

Saccharum spontaneum is also an important grass within the research area, particularly along water courses and, to a lesser extent, in high ridge areas. It is a perennial with "aggressive rhizomes" (Purseglove 1972:214) and tends to grow in dense clumps. However it is limited in distribution as it "requires a moister edaphic environment" (Ibid.:99).

Rottboellia sp. is particularly abundant in lower elevation areas and on ridges. Mercado (1978:11) remarks on its "competitive power" and its ability to spread rapidly. It is dispersed by wind, water and animals (Ibid.:36) and its seeds are viable for three years or more (Ibid.:20). Finally, it is not easily shaded out. In one test (Ibid.:23) *Rottboellia* was found to grow well in 80 percent shade, and still produce seeds in 73 percent shade.

The floristic composition of grasslands within the research area is very diverse and is conditioned by a number of factors. Paijmans (1976:100) notes that "The structure and composition of upper montane grassland are dependent mainly on altitude, soil depth, drainage conditions and . . . exposure."

The high elevation ridge-top grasslands, of which Kapisan (see map) is a good example, are dominated by three species--*Imperata cylindrica*, *Rottboellia sp.* and *Saccharum spontaneum*. Most of these grasses are somewhat stunted in such locations. Also very abundant is an unidentified pyrophytic fern. Scattered cycads may also be encountered. *Miscanthus* is usually present at the margins of these grasslands, though dwarfed or absent in exposed areas. Winds of high velocity are frequent here. The soils are heavily "laterized" oxisols, in contrast to the podzolic soils of neighboring mossy forest. Large-scale slump erosion is evident in these areas.

The Lomboy plain is one of the most interesting ecotypes within the research area. It is also one of the largest, running northwest for over four miles and covering nearly the entire southwest slope of Mt. Pinatubo up to the 1500 meter elevation. The Lomboy grassland is composed of *Imperata cylindrica*, *Rottboellia sp.*, some *Saccharum spontaneum* and two unidentified grasses (*polutput*, *laman puyo*). Only at the southwest end are a few small fire-resistant trees in evidence. This is a particularly impoverished floristic community and is almost completely devoid of herbaceous vegetation. Evidence suggests that it has been this way for a very long time. Smith (1909:23) states, "The coarse rocky soil at the base of Mt. Pinatubo contains practically no vegetation, the very small amount of grass which starts being periodically burned off by the Negritos." Photographs taken on the same expedition (Plate 7) confirm this description. The Spanish chronicler Perez, passing across the plain near Castillejos in the seventeenth century, an area geomorphologically and edaphically related to Lomboy, describes it thus: "The road is over sandy ground which is very large and full of rocks left by the river which flows from the mountain of Pinatuba [*sic*]; and in those places where there are no rocks, but only the sand, the road is also very wearisome because that sand has no cohesion, and the least wind that blows lifts the dust. . . ." (Blair and Robertson 1909, 47:296). It is conceivable that this is a naturally occurring grassland, for edaphic reasons. The "soil" and substrate of the Lomboy area is little more than sand, in excess of 50 meters thick, covered with coarse boulders. The water-holding capacity of this sand is negligible. As Richards (1952:315) notes, "edaphic factors are locally overcome or modify the effects of climate." It may then be that the Lomboy plain is what Walter (1971:230) would call an "edaphically conditioned grassland." Such an area might originally have supported a Molave type (*Vitex parviflora*) forest formation which, due to the gentle nature of the terrain and ease of access, may have been felled several centuries ago.

Miscanthus grassland is economically significant in that swiddens may be cleared from it. *Miscanthus* only occurs at higher elevations, reaching its fullest development above 600 meters in upland valleys and on protected slopes. The composition and density of these areas varies greatly. Commonly, it occurs in pure, nearly impenetrable stands. In other situations it may occur in clumps intermixed with *Imperata cylindrica*, *Musa sp.* and secondary tree species. Frequently *Chromolaena odorata* and *Rubus sp.* may dominate the herb layer.

Amukao

In 1953 Robert Fox noted that "The most characteristic flora of the entire Pinatubo area is the banana which occurs in tremendous stands" (1953:184). The endemic wild banana, *Musa sp.*,[20] deserves special attention as it is one of the most ubiquitous types of vegetation present within the research area. Standing 3-6 meters high, it exists gregariously in large, often nearly pure consociations, as well as isolated or in intermixed stands. Guido, writing about Ayta in the vicinity of Botolan, Zambales (1916:9), noted that:

> The most common kind [of banana] is the ones [*sic*] with seeds which the [*sic*] call "Amukao". . . . I have seen mountains, near Pinatubu Mountains [*sic*], with no plant on it, not even a weed, except for this kind of banana. I have walked for about a day through mountains planted with these bananas. They could not tell me, however, who planted these bananas. . . .

The distribution of *amukao* crosscuts nearly all vegetative zones, being found in areas ranging from primary montane forest to sun-baked grassland. It is among the most useful plants employed by the Ayta, being used for shelter, the preparation and serving of food, the

---

20. Probably *Musa errans* or *M. balbisiana*. The genus *Musa* however is more diverse than has generally been assumed (Lamoreaux, pers. comm.) and the *amukao* may not be one of these species.

construction of fishing dams, and many other things.[21] Productive swiddens can also be cleared from areas dominated by this species.

The Ayta recognize four varieties of wild bananas, only two of which exist within the research area; *amukao* and *alhang*. *Amukao* is by far the most common, being the variety that forms huge stands. *Alhang* is generally found in deeply forested areas, particularly along streams. It grows singly or in small clumps of only a few individuals. It is much smaller than *amukao*, seldom exceeding 2 meters in height, and has a psuedostem generally under 10 cm. in diameter.

Before widespread clearing occurred the wild banana presumably existed as a "nomad." Simmonds (1962:34) discussing the species *Musa balbisiana*, describes these as:

> opportunistic . . . ecologically seral plants. . . . The wild bananas are "jungle weeds" that spring up in transient, often man-made habitats; without human disturbance they would be rare plants confined to talus slopes, unstable hillsides liable to landslip or to forested areas devastated by periodical storms.

Amukao is adapted to a wide range of conditions, far outclassing most other Musaceae in this respect. As a whole the Musa family has a low tolerance of drought and shade, and they are poor competitors. Simmonds (1962:31) notes that *Musa balbisiana* is the most drought-resistant member of the family and that it "is perhaps more tolerant of grass competition than most" (Ibid.:32-33).

Wild bananas are dispersed primarily through the activities of birds, bats and other small animals. The fruits are full of small black seeds surrounded by pulp. Simmonds (1962:1) remarks that:

> The pulp may be thought of as an adaptation to animal distribution for presumably it is this rather than the seeds which is sought by the monkeys, bats and birds that always attack bunches that start to ripen in the field.

---

21. Interestingly, Brown and Mathews (1914) report that it is absent from the eastern slopes of Bataan, only 35-40 miles south of the research area.

Once germinated and established, *amukao* is extremely durable. Simmonds (1962:2) notes that, "Though individual stems die after flowering, the plants are potentially immortal." They are not killed or otherwise negatively affected by burning: individual pseudostems may be destroyed, but vigorous sprouting soon occurs. Such burning may in fact stimulate further germination. According to Simmonds (Ibid.:34):

> Buried, viable seeds germinate when the vegetation and soil are subjected to various . . . disturbances such as fire, clearing, cultivation, landslip, etc.; only after severe disturbance does one ever see locally abundant germination in the field and then it may be very abundant indeed.

In terms of the floristic composition of *amukao* stands, the species is found in a great variety of edaphic situations in association with many other types of plants. Frequently, at higher elevations, it is found in association with *Miscanthus* and small secondary tree species. Often it occurs in nearly pure stands with only a very sparse herb layer: in such cases *Chromolaena odorata*, *Hyptis sp*. and several species of small grasses are the most characteristic plants.

Riparian vegetation

The areas bordering streams and rivers support an anomalous and rather variable mixed assemblage of plants, a category in which species characteristic of both primary and secondary associations may be found. Riparian vegetation differs from other communities not so much in terms of distinct taxa, but rather in the way taxa characteristic of several biotic zones are combined.

The zone bordering streams is considerably wetter than in adjacent areas, thus largely ameliorating the effects of seasonal drought. Edaphically these areas range from sandy, seasonally inundated river beds, to river terraces, to various combinations of inseptisols, ultisols and alfisols and, where widespread seepage occurs, to azonal podzols. The water balance of these areas is considerably different from that of surrounding slopes. In some cases, the water table may be very near the surface. Also, where these environments occur in deep valleys or canyons, insolation is reduced considerably. Gully forests may further provide insulation. Humidity is thus locally much higher, allowing the existence of mesophytes.

The most distinct riparian vegetative association is found in *ka'ogan* areas. A *ka'ogan* is a wide, flat, sandy river bed. Only one exists in the research area; the upper reaches of the Santo Tomas river at the foot of the mountains, paralleling the western boundary of the research area. This river bed exceeds 200 meters in width at some points, becoming smaller as one moves upstream toward Lomboy. The most distinctive biotic feature here is the many *Casuarina equistafolia* trees. Except for the lowest reaches of a few second-order streams which feed into it, and along the streams bordering the Lomboy plain, this is the only area that supports *Casuarina*. Clumps of *Saccharum spontaneum* are also extremely common in *ka'ogan* areas. Fox (1953:184) notes that the shrub *Homonia riparia* and the two species of grass *Phragmites vulgaris* and *P. karka* are common in such areas as well. *Ka'ogan* areas are very open and subject to extreme isolation.

The slopes surrounding most streams within the research area have been largely deforested. Usually however "gully forest" (*sensu* Robbins 1968:529) remains. In such areas tall emergents such as *Dipterocarpus vernicifluus* and *D. grandiflorus* may remain. Other large trees, though scattered, are *Chisocheton cumingianus*, *Pterocarpus sp.*, *Toona sp.*, *Palaquium spp.*, and the palm *Caryota cumingii*.

What is most distinctive about riparian vegetation is the heavy development of the ground and field layers. In areas more exposed to light, numerous vines and the scrambling bamboo *Dinacloa sp.* are common. *Thysanolaena maxima* is also found here. In areas where the canopy is more dense, lithophytic and epiphytic plants, especially ferns, increase dramatically in abundance; *Asplenium nidus* is most conspicuous. The ground orchid *Spathoglottis sp.* is also present, as may be *Freycinetia sp.* and other plants characteristic of moister (and in this case, higher elevation) edaphic environments (Whitford 1906:651). The vegetation along streams in higher elevation montane and mossy forest remains largely undisturbed. It appears to be of largely the same species composition as forest on adjacent slopes, though detailed survey may show otherwise. In such areas large numbers of tree ferns (*Cyathea spp.*) and lone *Musa sp*. are found.

CHAPTER 6

SWIDDENING

A. The Study of Swidden Systems

The previous three chapters have considered the context in which disturbance in upland Zambales occurs. It is now possible to examine these disturbances themselves. Of the three classes to be considered--swiddening, burning and logging--I discuss swiddening first and treat it in the greatest detail. This is because this activity has been the primary form of initial forest clearance in upland Zambales for the longest period of time and continues to be the dominant form today. It is the most purposive, complex and variable form as well, the foundation upon which Ayta participation in the dynamics of this upland ecosystem rests. Succession theory is a particularly useful approach to the study of swidden systems. Shifting cultivators, by clearing land, usually in an annual cycle, affect the environment in a manner which influences its ecosystemic properties--the cycling of nutrients and energy--as well as species composition.

Before turning more specifically to the Ayta swidden system, it may be useful to provide a brief historical review of trends in the study of swidden systems. Studies of societies practicing swidden agriculture have played a central role in the development of contemporary ecological anthropology. Many of the principles employed today were developed in the context of such societies. Taking cues from geography and agronomy, the traditional view of swidden agriculture was largely negative. Swiddening was generally held to be wasteful, destructive and inefficient, and swiddeners themselves were portrayed as being lazy, bound by tradition and unconcerned about the impact of their activities on the environment. An evolutionary interpretation was often imposed, the view being that shifting cultivation represented an archaic form of agriculture. Gourou (1956:339), for instance, stated that "As with all agricultural techniques, *ladang*[22] is the expression of a civilizational level and not the result of any inescapable physical constraint." Elsewhere (1958:49) he states that "the *ladang* system . . . hinders the rational management of the agricultural land . . . [it] invests nothing in the soil and this inevitably results in low yields." Further, "The *ladang* system is at [*sic*] the final analysis an inadequate economic basis on which a high civilization may achieve great political and intellectual attainments" (Ibid.:52) Sutherland (1971, quoted by Janzen 1973:1216), perhaps sums up this attitude best: "What is wrong with subsistence agriculture is that everything that is produced is used up by the people." The logical extension of such assumptions was that swiddeners could, and certainly should, progress. This implied the abandonment of traditional subsistence systems and the adoption of permanent, fixed-field agriculture.

In the mid-1950s this view began to change. Several influential studies appeared which called the negative assumptions such as those above into question. Perhaps the most important was Conklin's 1954 article "An Ethnoecological Approach to Shifting Cultivation." In this brief article, Conklin described a number of the then current misconceptions about swidden agriculture, and set out a ten point corrective which has since become the foundation of most contemporary swidden studies. This was followed three years later, in 1957, by "Hanunoo Agriculture." In this he reinforced his original ten points by distinguishing between "partial" and "integral" swidden systems. Partial systems were those in which the "farmers have strong sociocultural ties outside the immediate swidden area into which they bring permanent-field agricultural concepts" (1957:3). In general partial swiddeners are not familiar with swidden techniques and are sloppy or negligent in their methods. Integral systems on the other hand are the result of many generations of observation and fine-tuning and are, therefore, sustainable. Integral swiddeners know what they are doing, understand ecosystemic processes, and recognize the consequences of their actions.

22. *Ladang* is the term for a cultivated field or farm in Indonesia and Malaysia, and is the general term applied to swidden farming.

Though Conklin's distinction between partial and integral systems was useful in reinforcing the fact that swiddening had destructive potential if not done properly, the interpretation cultivated by most anthropologists thereafter was that traditional swidden systems lacked destructive potential and were innately sustainable. This view was reinforced by the 1963 publication of Geertz's *Agricultural Involution*. In his essay "Two Types of Ecosystems," he compared swiddening with intensive wet rice cultivation. Swiddening, "when genuinely adaptive, maintains the general structure of the preexisting natural ecosystem into which it is projected" (1963:16). "A natural forest has been transformed into a harvestable forest" (Ibid.:25). This has become the standard view on the adaptability of swidden systems. Moran (1979:270), for instance, states "The practice of slash-and-burn agriculture is not only economically sensible, but is also ecologically sound. . . . The intercropped fields of swidden horticulturalists are an imitation of the natural ecosystem; generalized and diverse, multistoried, and providing good soil protection from both solar radiation and precipitation."

It is against this background that most modern swidden studies have been undertaken. The literature supporting this perspective has grown steadily.[23] Without question, the view put forward by Conklin and Geertz has been valuable insofar as it serves to bring to our attention the sophistication of traditional swidden systems, allowing observers to recognize the detailed ecological knowledge possessed by practitioners of such methods. Be this as it may, the view taken here is that, while the older perspective of Gourou and others is clearly in error, there are also certain problems with the newer view. Both make erroneous *a priori* assumptions, albeit of a very different sort. It is certainly unfair to place the burden of guilt on Conklin and Geertz for any shortcomings in the more recent view. Both were careful to qualify their statements. Conklin in particular exemplified a high degree of ecological exactitude and acumen. Rather, much of the burden of guilt should be placed on those who, ignoring contrary evidence, have continued to uncritically accept the notion that traditional swidden systems are intrinsically sustainable. Most often they indeed are. But as a study orientation, such an assumption is decidedly unsatisfactory as a guide to understanding swidden systems.

---

23. See, for example, Vayda and Rappaport 1968, and Netting 1977.

Perhaps part of the reason this assumption has persisted has to do with the emphasis on homeostasis and equilibrium discussed in Chapter 1. This emphasis has carried over into studies of swidden societies. Bennet, for instance (1973:38), citing Street (1969), notes that anthropologists have not tried:

> to determine whether degradation of the soil and land surface was in fact occurring. Street remarks that since anthropologists tended to assume that agricultural practices of swidden peoples were permanent and unvarying, they therefore did not examine methods of resource renewal or conservation.

When degradation has been considered it is often in the context of "intensification," usually involving higher-level evolutionary implications. An oversimplified, non-ecosystemic equation has therefore been assumed: population pressure begets intensification begets shortened fallow begets degradation. This type of thinking has resulted in the creation of concepts such as "carrying capacity." Systems were held to be stable up to a certain point at which, suddenly, carrying capacity was exceeded, and at which point they would break down. What has been ignored is the great number of factors impinging on the sustainability of particular swidden systems. Neither the stability nor instability of particular swidden systems can be taken as given. Rather, this is dependent on a wide range of factors which must be specified. Succession theory allows us to specify these.

B. The Ayta Swidden System

Ayta in Zambales are today firmly and fully committed to a subsistence system based on swidden agriculture. Evidence suggests that such a reliance has been of long standing. Today nearly 100 percent of the total diet is derived from swiddening.

A large crop complex with numerous varieties of specific cultivars is planted by Ayta in the Cabusilan area. The primary cultigens are sweet potatoes (*kamoti*; *Ipomoea batatas*) and *Xanthosoma* (*biga*). Of lesser importance are corn (*ma-i*; *Zea mays*), bananas (*ha'a*; *Musa sapientum, M. paradisiaca*), taro (*talu*; *Colocasia esculenta*) and rice (*pali*; *Oryza sativa*). Pigeon pea (*kade*; *Cajanus cajan*) and lima bean (*patani*; *Phaseolus lunatus*) are important legumes and, when available during the dry season, may comprise 10 percent of the diet by weight. Cassava (*dotong, moros*; *Manihot esculenta*), yams (*ubi*; *Dioscorea alata*), and squash (*kalubaha*; *Cucurbita maxima*) are also present, but are generally of only peripheral significance. *Bagoyboy* (*Thysalolaena maxima*), an endemic grass, may also be planted. The panicles are traded with lowlanders, who use them in the manufacture of brooms. Numerous varieties of all the major cultigens are recognized.

It appears that swiddening has been a significant component of Ayta subsistence for a very long time. Fox (1953:247) states that "The association of the Negritos with cultivated plants must be reckoned in a few hundred years--excepting perhaps the taro and yams. . . ."[24] It is quite conceivable that Ayta in Zambales have been relying on swidden agriculture since prior to the arrival of the Spanish. Spencer (1966:36) has stated that, "As described in 1904, the Negritos of Zambales, western Luzon Island, cannot long have been growing crops, nor can they have ranked very high in the scale as crop growers." Such a statement does not seem warranted. Several types of evidence point to the antiquity of swiddening by Ayta in upland Zambales. First there is the significant degree of deforestation which has occurred. While some of this has been due to logging in recent decades, most can be accounted for primarily with reference to swiddening. A second line of evidence is the large crop complex which has been assembled by the Ayta. Multiple varieties of many cultigens are recognized and grown. Such a rich inventory of crops, and the detailed knowledge necessary to successfully grow these, requires some time to establish. It appears that this crop complex was fully in place by at least the beginning of the present century, and likely much earlier.

The third and most compelling line of evidence for the antiquity of swiddening in this area is the historical record. These sources are not always reliable, particularly some of the older accounts. Spanish priests and administrators invariably characterized Negritos as naked, wild, and certainly incapable of anything

---

24. To this might be added bananas.

approaching agriculture. Despite this, comments to the contrary are found in historical documents. Not all these accounts point clearly or directly to the Ayta, but indicate that swiddening was being practiced by unspecified upland groups in the vicinity. Until the early 19th century there were Sambal inhabiting certain upland areas in Zambales Province, particularly to the north and west of the research area.

Loarca, speaking of mountain-dwelling Sambal in 1582, notes simply that "They cultivate the land although only in small tracts" (Blair and Robertson 1909, 5:105). Speaking of Ayta in the Zambales/Pangasinan area in 1618, one account states contradictorily that "Very few of them have fixed settlements, nor do they plant crops; but they live upon camotes (a kind of potato), other herbs and roots, and the game which they hunt" (Ibid., 18:98). Gemelli Careri (1963:39) reported in 1696 that:

> The inland is full of wild indians, who like brute beasts, wander naked up and down the woods and mountains, only covering their privities with a leaf. They sow a little in their valleys, and what more they want, they get in conquered country. . . .

More recent accounts, from about the turn of the century, are considerably more accurate. Parker (1964:116) in 1908, speaks of the "mountain maize" grown by Ayta, and remarks on "The mountain rice that can be raised in large quantities and that is much superior to valley rice. . . ." Reed's remarks (1904:42) are worth quoting at some length:

> By far the greater part of the Negrito's energies are directed to the growing of tobacco, maize, and vegetables. He does not plant rice to any extent. All planting is done in cleared spots in the forest, because the soil is loose and needs no plowing as in the case of the lowland. The small trees and underbrush are cut away and burned and the large trees are killed, for the Negrito has learned the two important things in primitive farming--first, that the crops will not thrive in the shade, and second, that a tree too large to cut may

be killed by cutting a ring around it to prevent the flow of sap. The clearings are never large.

Usually each family has its clearing in a separate place, though sometimes two or more families may cultivate adjoining clearings. The places are selected with a view to richness of soil and ease in clearing. In addition to preparing the ground it is necessary to build a fence around the clearing in order to keep out wild hogs.

Informants state that "before," certainly sometime before the present century, their swiddens were smaller. They claim to have had only *kulub*, the Ayta word for a small swidden, and lacked *gitna* or *ginawa*, the full-sized swidden of today, which averages .2-.25 hectares. This suggests a progressively greater reliance on swiddening within historical memory but, significantly, Ayta make no reference to hunting and gathering when recalling their past, other than in the most vague terms. It will be recalled that the Ayta do speak of a time when their ancestors were hunter-gatherers, but this is exceedingly vague and in an almost mythical far distant past.[25]

One uncertain and admittedly speculative piece of evidence for the antiquity of swiddening by Ayta relates to the size of sub-dialect areas. Four or five sub-dialects are presently spoken by Ayta within the Zambales area, each with perhaps 1000 speakers, in areas of approximately 100 sq. kilometers. In comparison to other Negrito groups in the Philippines who are still actively engaged in hunting and gathering (see Estioko and Griffin 1975, 1981), this represents a

25. As noted in Chapter 2, Ayta historical memory in the matter of subsistence is very different from that of former hunter-gatherers in Sarawak, East Malaysia. Groups such as Penan Suai and Penan Jelalong, who have been settled agriculturalists for close to 200 years, nevertheless retain a vivid and unambiguous oral historical tradition of having been hunter-gatherers, even to the extent of being able to specify which of their ancestors was the first to plant crops. Ayta accounts, by contrast, are much more vague, perhaps suggesting that, for them, the transition from hunting and gathering to agriculture occurred in a more distant past.

considerable constriction of subdialect area size, though interestingly the population sizes concerned are closely approximate. This may suggest that subdialectical differences have evolved with the increasing sedentarization and territorial identification coincident with greater reliance on swiddening.

One feature of the present Ayta swidden system is the overwhelming dominance of New World crops. Fox (1953:175) noted that "All [Ayta] desire their basic subsistence from three introduced New World crops: the sweet potato, the cassava, and corn." Over 90 percent of the present diet is comprised of New World crops. Sweet potato and *Xanthosoma* are the dominant cultigens, together comprising over 80 percent of the diet. This estimate differs somewhat from Fox's figures (1953:246), though perhaps not significantly. He estimates that "approximately 70 percent (!) of the Pinatubo pygmies' present food is derived from plants introduced during the Historic Period from the New World." His estimate of sweet potato composing 53 percent of the diet (Ibid.:246) is borne out by my own observations. It is interesting, however, that nowhere does Fox mention the presence of *Xanthosoma*.[26] The legumes lima bean and pigeon pea, both of which are introduced, comprise another 10 percent of the diet.[27] Bananas, taro, cassava, and rice together comprise 5-10 percent of the remaining diet, depending upon location in the research area. Kalantas, for instance, has considerably more bananas than Maagə'agə'.

Clearly a major subsistence transformation has occurred within a post-Hispanic time frame, through both a replacement of indigenous crops and through extensification, i.e., an increase in the scale of swiddening, and thus fallow areas, over a greater percentage of the landscape. This suggestion has two possible correlates: (1) that swiddening, if practiced in pre-Spanish times, would have been based primarily on Old World crops such as bananas, taro, and yams, and (2) population expansion has occurred. The present Ayta population

---

26. Though this may be due to the fact that Fox worked in a different area, to the northwest of the present research area, it is still peculiar that he does not mention *Xanthosoma*.

27. Lima beans are a New World crop, while pigeon pea originates in Africa. Along with *Xanthosoma*, Fox also makes no mention of pigeon pea.

density is approximately 10 per sq. kilometer, a density highly uncharacteristic of hunting-gathering populations, but well within the limits of expectation for tropical swiddeners. New World crops with their higher levels of productivity have to some extent made this possible.

C. Contemporary Cropping Patterns

The present swidden system is the product of a long historical sequence. It has been shaped both by factors of agronomic necessity and of human preference. The following is an analysis of this system, based upon a sample of 39 swiddens which were observed in the course of research. I estimate that this sample represents 40-50 percent of the total number of swiddens within the research area.

In looking at particular swiddens, data were collected on a range of factors, most of these having relevance to successional considerations. Fourteen classes of data were collected for each swidden. These included:

(1) Owner
(2) Location
(3) Age of swidden
(4) Crops present
(5) Dates cleared
(6) Date burned
(7) Dates planted
(8) Previous vegetation
(9) Surrounding vegetation
(10) Elevation
(11) Degree of slope
(12) Aspect
(13) Weeds present
(14) Spacing of crops

Additionally, for each swidden I collected data on such things as the varieties of cultigens present, whether coppiced trees were killed or not, and recorded the presence of conspicuous features such as rock outcrops. Two types of information are notably lacking: (1) data on soils, and (2) production data on specific swiddens. The latter is particularly difficult to collect, as each family possesses several swiddens which are visited in no particular order over the period of a year or more. There is no single harvest which can be measured other than that of rice. Also a great deal of "borrowing" occurs among extended family members from different settlements, and it requires a tremendous amount of effort to keep track of which swiddens various food items are coming from.

As a supplement to this survey information, several classes of data were subjected to statistical procedures using SPSS (Nie et al. 1975), including subprograms FREQUENCIES, CROSSTABS, and NONPAR CORR. The completed format is presented in Appendix A. For purposes of the analysis, eleven cultigens were chosen for inclusion. These eleven are either staples, of particular seasonal importance, or have economic significance. Less important cultivars which occur in small numbers and have little ecological or nutritional significance are excluded. Nine classes of vegetational community (PREVEG), both primary and secondary, were also included in this statistical analysis. It will be noticed that the vegetational categories used in this analysis differ from those presented in Chapter 5. Primary forest types have been aggregated while various types of secondary vegetation have been disaggregated and/or recombined. This was found to be necessary in order to more accurately reflect the range of vegetational associations from which swiddens are cleared, and which may more accurately impinge upon their productivity and successional trajectories.

Each household generally clears at least two swidden plots every year, some as many as three. These swiddens are relatively small, ranging in size from .15 hectares to 1.5 hectares, the majority being from .2 to .25 hectares.[28] Swiddens are frequently cleared on slopes of extreme steepness, the maximum slope generally between 30° and 55°, with slopes of 60° and greater having been observed. Ridges and outcrops are a favorite site for swidden clearance. As Conklin (1959) has suggested, this may contribute to the establishment of grasslands.

---

28. Garvan (1963:77), referring to Ayta in the first decades of the 20th century, states that "Small clearings, up to half a hectare in area, are made every year. . . ."

Ayta within the research area tend to inhabit higher elevations. Settlements range in elevation from 200 meters to 1000 meters with the majority above 500 meters. This is reflected in the placement of swiddens. Within the sample of 39 swiddens, four were between 400 and 500 meters, fifteen between 550 and 800 meters, and the remaining twenty were from 900 to 1020 meters. This is in striking contradiction to Fox, who states (1953:183) that even the "'deep', more isolated, groups living on the slopes of Mt. Pinatubo" are "never higher than 1,500'."[29]

The Ayta swidden cycle is a function of the annual cycle of wet and dry seasons. The clearing of swiddens generally begins in early February and runs through the first part of April, though there may be clearings made before or after this. March is considered the best time to clear, and this is when most such activity takes place. The clearing of swiddens may be the work of a single family, or may be done cooperatively by large groups. Cleared swiddens are left to dry for one or two months. Most burning takes place toward the end of April. Days with a light, steady breeze are considered optimal, and on such days many swiddens may be burned. Interestingly, escape fires do not appear to be a serious problem; none were directly observed, and aerial photographs indicate only one possible case.

Immediately after burning a swidden, often while it is still smoldering, corn is planted. If the swidden has been cleared from *amukao*, the pseudostems, which become mushy and do not burn, are collected and stacked in piles throughout the swidden. Weeds are then thrown on top to decompose. Once the rains begin these piles (*ibonton*) provide edible mushrooms.

From mid-May to mid-June rice may be interplanted with the corn which was planted earlier. The timing of this depends upon the prospects for rain and the availability of labor. Rice planting is always a collective effort. This corn/rice crop combination is generally the case only in lower- to mid-elevation swiddens. Rice was never observed above 800 meters. Crops such as *Xanthosoma*, taro, sweet potato, pigeon pea and lima bean are generally planted from the end of May through the month of June. In any particular swidden,

---

29. On a survey flight over the area where Fox worked, Ayta swiddens and settlements were spotted at elevations of over 750 m (2460'), on the middle slopes of Mt. Pinatubo.

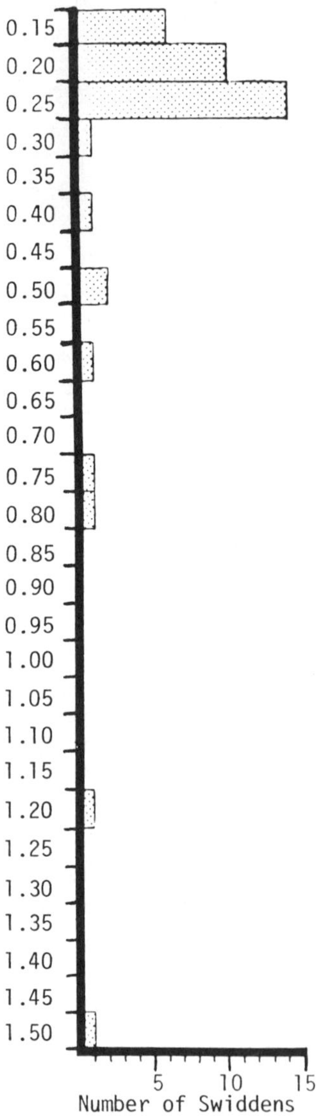

Figure 6. Distribution of Swiddens by Size (in Hectares).

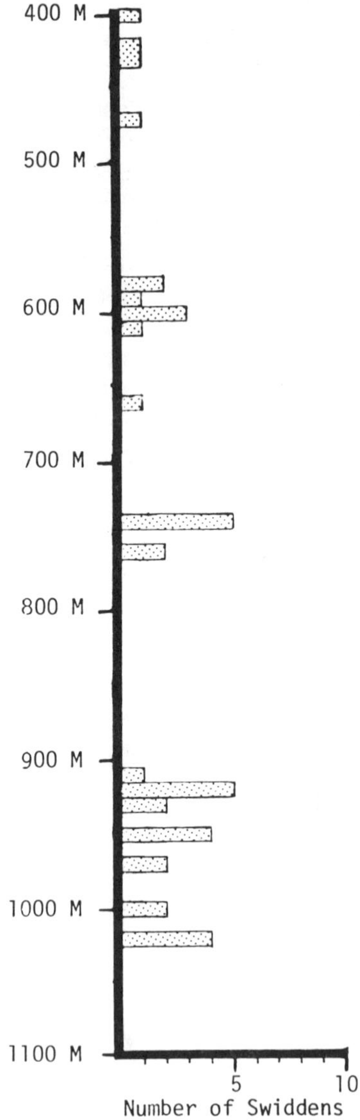

Figure 7. Distribution of Swiddens by Elevation.

planting may be a continuous, drawn-out process over a period of several weeks, particularly in the case of root crops which are vegetatively propagated.

The process of weeding consumes a large amount of time: each swidden must be weeded at least twice during the cropping cycle. The first weeding generally occurs about a month after planting. This is frequently a communal effort in which several households may participate. The timing of the second weeding is more variable. Sporadic weeding is also done as needed.

By the middle of June the first corn becomes available and by September rice can generally be harvested. Very low yields are usually realized from these two crops and, particularly with rice, destruction by rats may occur. Root crops take slightly longer to mature. The first sweet potato and *Xanthosoma* can be harvested by October. By December, when the dry season has set in, it is possible to harvest pigeon pea and lima bean, and these produce throughout the dry season.

In the 600 to 800 meter range, where both rice and sweet potatoes occur, sweet potato may be planted following the harvest of rice, so as to become established before the dry months commence. Additional legumes may also be planted during this time.

Overall, the yields derived from swiddening are quite high whereas, though my data in this respect are incomplete, labor inputs appear relatively low. The Ayta do not suffer from any shortage of food *per se*. I was in fact impressed by the sheer unvarying volume of the diet. At least twice a day large meals are prepared. Large piles of sweet potatoes, *Xanthosoma* or some other carbohydrate are served on banana leaves. Furthermore there appears to be little in the way of seasonal fluctuation in the overall availability of food, and informants state that, with the exception of World War II, periods of hunger have never occurred. The nature of the diet is another matter. As noted, carbohydrates, primarily in the form of sweet potato and *Xanthosoma*, comprise by far the greater part of the diet. Meat and fat are conspicuously absent.

Elevation was found to be a very important variable in determining the occurrence of specific crops. In general those crops that are most important in terms of their total contribution to the Ayta diet increase in occurrence with elevation. Sweet potato and *Xanthosoma*, which comprise in excess of 80 percent of the total diet,

	400 M	420 M	430 M	470 M	580 M	590 M	600 M	610 M	660 M	740 M	760 M	910 M	920 M	930 M	935 M	950 M	970 M	1000 M	1020 M
Sweet potato						1	1	1	1	2	2	1	5	1	1	4	2	2	4
Xanthosoma		1	1		2		2	1	1	4	1	1	4	1	1	3	2	2	3
Corn	1	1	1	1	2		2	1	1	3	1		4	1		1	1		
Colocasia		1			1			1		1		1	4	1		1	2	2	2
Rice	1	1	1	1			2	1		1	1								
Banana		1							1	2	1		1						
Cassava				1		1	1	1			1					1	1		
Ubi		1	1		1			1	1	2	1		1					1	
Pigeon pea		1	1	1	2		2	1		1	1		3			1	2	1	1
Lima bean	1		1		2		2	1		2	1		3			1	2	1	1
Bagoyboy		1	1	1	1	1	1				1	1			1	1	1		1
Total swiddens	1	1	1	1	2	1	3	1	1	5	2	1	5	1	1	4	2	2	4

Table 1. Crops by Elevation

are grown at the highest elevations. In the sample of swiddens observed, sweet potato was completely absent below 590 meters and was present in all swiddens above 760 meters. The pattern for *Xanthosoma* was similar, though not as extreme. In only two of thirty cases was it found below 600 meters, and was absent in only three of the twenty swiddens above 910 meters. Taro, though not as important as the other two crops, followed a similar pattern. Two of the seventeen occurrences of taro were below 600 meters, and most were found above 900 meters.

At the opposite extreme, corn, cassava and rice are most important at lower elevations, though the tendency is not as strong as in higher elevation crops. Corn was present in ten of twelve swiddens 660 meters or less, though it was also present in a few swiddens above 930 meters. Cassava was present in five of seven swiddens below 760 meters, though it was also present in two cases at 950 meters and 970 meters. Rice was present in all four swiddens below 500 meters and was very prevalent in swiddens between 600 meters and 750 meters, but it was absent above this.

Those crops showing wide variability with respect to the elevation at which they occur are pigeon pea, lima bean and *bagoyboy*. Banana is interesting in that all but one case they were clustered between 660 meters and 760 meters. This may in part be a function of the sample, as most bananas in my sample were found in Kalantas, which is within this elevation range.

The direction in which a swidden faces was found to have no predictive value in terms of crops present. It should be noted, however, that swiddens clustered somewhat between 135°-180° and 270°-360°. This is due to sampling bias rather than to any particular factor.

Degree of inclination was found to be significant in the case of several cultigens. Taro was found to be cultivated on steeper slopes and none was present in swiddens with a maximum slope of under 25°. This is due to the practice of planting taro at the bottoms of swiddens, frequently adjacent to streams, where they are shaded and where soils are usually wetter. Such areas tend to be extremely steep. A similar tendency was apparent in the case of pigeon pea, though the reason for this is not clear. Fifteen of eighteen swiddens in which pigeon pea was present had maximum slopes in excess of 30°. *Bagoyboy* also occurred in steeper swiddens, those with a maximum slope of 25°-50°. Only one case deviated from this.

Several cultigens were generally found in more level swiddens. Most marked was cassava; minimum slope in nearly all cases was less than 20° and maximum slope never exceeded 40°. *Ubi* also tended to occur in more level swiddens. Seven of ten swiddens in which it occurred had a minimum slope of under 10°. All other crops demonstrated a wide range of variability in the steepness of swiddens in which they occurred.

A strong relationship was found to exist between swidden size and elevation, with a trend toward decreasing size with increasing elevation. Five of the seven swiddens .4 hectares or larger were below 610 meters. The other two were at 950 meters and 1000 meters. Most of the very small swiddens (1.5-2.0 hectares) were above 740 meters. Only two of sixteen in this class deviate. Very high levels of significance were found in the crosstabulation of these variables. Cramers-V was .79383, contingency coefficient was .929 and Eta was .81968 with size dependent. It is interesting that the large number of swiddens .25 hectares in size are spread very evenly by elevation at all but the very lowest elevations.

There is a sharp dichotomy between larger and smaller swiddens. Thirty swiddens (77%) were within the .15-.25 hectare size range while only nine swiddens (23%) were within the .3-1.5 hectare size range.

The relationship between specific crops and swidden size was examined, these crops compared, and the extent to which size correlations are actually expressed as crop complexes was considered.

Sweet potato: Concentrated in small swiddens. Of the 28 swiddens in which sweet potato is present, 25 are in swiddens .25 hectares or less.

Xanthosoma: Variable.

Corn: Codominant in 7 of 9 swiddens .3 hectares or greater in size. Present, though with high scores, in only 14 of 30 swiddens .25 hectares or less.

Taro: Very few in large swiddens. Almost exclusively in swiddens .25 hectares or less.

Rice: Present in 6 of 8 swiddens .3 hectares or larger, and in only 2 of 30 swiddens .25 hectares or less.

Banana: Concentrated in smallest swiddens, though also present in the largest swidden in sample.

Cassava: Variable, with bias toward larger swiddens.

Yam: Variable, though concentrated on both ends of scale. 7 in swiddens .25 hectares or less, 3 in swiddens .75 hectares or greater.

Pigeon pea: Variable.

Lima bean: Variable.

Bagoyboy: Variable.

Sweet potato, taro and banana were clearly more prevalent in smaller swiddens, while corn, rice and to a lesser extent cassava occur primarily in large swiddens. *Xanthosoma*, yams. pigeon pea, lima bean and *bagoyboy* exhibit wide variability.

In terms of crop complexes, as opposed to individual cultigens, reference to the Spearman correlation coefficient matrix (Appendix B) shows some interesting relationships. Sweet potato shows a strong opposition to both corn and rice, with correlation coefficients of -0.6482 and 0.6005, respectively, both with a significance level of .001. As all these tend to have high values when present, there appears to be a strong dichotomy, which is associated with swidden size.

No significant clustering of Old vs. New World cultigens with different factors was found. However, if relative importance in the diet is considered it is clear that the two primary cultigens, sweet potato and *Xanthosoma*, increase in importance with elevation.

Certain New World and Old World crops show a very high level of correspondence, particularly those belonging to the same family. Graminaceae, represented by rice and corn, and Araceae,

	Size (in Hectares)										
	0.15	0.20	0.25	0.30	0.40	0.50	0.60	0.75	0.80	1.20	1.50
Sweet potato	5	9	11		1		1	1			
Xanthosoma	4	8	12		1	1	1	1	1		1
Corn	4	2	8	1		2		1	1	1	1
Colocasia	3	4	7				1	1			1
Rice			2	1		2		1	1	1	1
Banana	1	2	2								1
Cassava		2	2	1	1			1			
Ubi		2	5					1	1		1
Pigeon pea	2	6	4	1	1	1		1	1		1
Lima bean	2	6	4		1	2	1	1	1		
Bagoyboy	1	3	4	1	1				1		1
Total swiddens	6	10	14	1	1	2	1	1	1	1	1

Table 2. Crops by Size

represented by *Xanthosoma* and *Colocasia* are cases in point. This is also true with the two primary legumes, pigeon pea and lima bean, though both are introduced. Several New World crops, primarily *Xanthosoma* and corn, fit into Old World crop niches and tend to reduce Old World cultigens to only peripheral significance.

The basic Ayta cropping pattern in the Cabusilan area then, consists of sweet potato occurring in higher elevation swiddens cleared from primary or mature secondary forest vegetation, with corn and rice at larger lower elevation swiddens cleared from earlier successional vegetation. Other crops of importance, particularly *Xanthosoma*, pigeon pea and lima bean are much more variable in their requirements and thus in the range of contexts in which they are found.

D. Cropping Patterns and Vegetational Associations

The mosaic of vegetational associations characteristic of the Cabusilan area have been of great significance in allowing continued habitation of these uplands. For an area so conspicuously deforested and degraded, a fairly dense human population is presently being supported. This is largely because it is possible to clear swiddens from a variety of types of vegetational association, both primary and secondary. The following discussion will consider this relationship--between particular types of vegetational association and the swiddens which are cleared from them.

The type of vegetation characteristic of a particular site was found to be an important determinant of swidden size. In general, swiddens cleared from earlier successional vegetation tend to be larger, whereas those cleared from mature secondary and climax vegetation are considerably smaller. The relation between various vegetational associations and swidden size can be summarized as follows:

Grassland: Associated only with large swiddens; two cases--.8 hectares and 1.5 hectares.

Miscanthus: Associated with swiddens of all sizes, though with some bias toward larger swiddens.

Amukao: Associated with larger swiddens.

Young secondary forest: Variable, with some bias toward larger swiddens.

Miscanthus/amukao: Variable.

Miscanthus/amukao/young secondary forest: Associated with swiddens of all sizes, but more strongly associated with smaller swiddens (11 of 13 cases).

Mature secondary forest: Associated with smaller swiddens, and in no swidden larger than .5 hectares.

Primary forest: Exclusively associated with smaller swiddens, none larger than .5 hectares.

Riparian vegetation: Variable.

Several trends were observed in the relationship between specific crops and specific vegetational associations. These are summarized in Appendix B. Sweet potato was an interesting case. The only PREVEG it correlated with positively with any degree of significance was primary forest. It had a fairly marked negative correlation with all earlier successional vegetation and with riparian vegetation. *Miscanthus/amukao* and *Miscanthus/amukao*/young secondary forest all showed low correlation coefficients, which implies their generalization. Rice was at the opposite extreme, showing a very high association with early successional vegetation. Corn surprisingly showed less positive correlation coefficients with the same types of PREVEG. *Xanthosoma* was found to occur much the same as sweet potato, the primary difference being that it was strongly associated with mature secondary vegetation rather than primary vegetation. A very strong association was evident between *amukao* and young secondary forest. *Colocasia* did not correlate strongly positive with any type of association and only moderately negative with *amukao* and young secondary forest. It was slightly positive with *Miscanthus*.

Pigeon pea and lima bean were shown to be very generalized. They were negatively correlated only with primary forest.

Several crop complexes have been identified and in two cases associated with specific vegetative associations. These can be reviewed with reference to elevation and inclination. It has been demonstrated that aspect is not predictive.

Sweet potato/primary forest: Both occur at highest elevations. Sweet potato is found on wide range of inclinations, primary forest biased toward steepness.

Corn/rice/early successional vegetation: All occur most heavily at lower elevations and on variable slopes.

Xanthosoma/*Colocasia*: Generalized in terms of PREVEG, though both show some association with riparian vegetation. *Colocasia* tends toward steep inclination, *Xanthosoma* is variable. Both are negatively correlated with *amukao* and young secondary forest.

Pigeon pea/lima bean: Generalized in terms of PREVEG. Both variable in terms of elevation and inclination.

Cassava/*bagoyboy*: Generalists in terms of PREVEG, but do not occur with primary forest.

Whereas certain crops show strong correlation to specific classes of PREVEG, elevation and inclination, others are more generalized, cross-cutting vegetation types and other factors. Sweet potato, corn and rice are examples of the former, while *Xanthosoma*, pigeon pea and lima bean are examples of the latter.

E. Parameters of Disturbance: Frequency

When referring to the length of fallow period for a particular swidden site, we are in fact talking about the frequency with which that site undergoes disturbance. In areas where primary montane forest is being cleared, frequency of disturbance is, of course, an irrelevant parameter. However, the current amount of clearance occurring in primary forest is small. I do feel that there is a bias in my sample indicating a higher degree of such clearing than is generally the case. Bayabas appears to be one of the few places where this is taking place on any scale.

The length of time deemed appropriate to leave an area under fallow varies, depending upon the nature of the vegetation which has become established on a site. For instance, an area in which *Miscanthus* or *amukao* are dominant may be cleared within five to eight years. Areas which develop a woody fallow early in succession (in contrast to woody fallows which follow an intermediate grass or *amukao* state) require somewhat longer, approximately ten years. These figures refer to the minimum fallow period. Many areas are not cleared this soon and may be passed over for many years. Thus they continue to undergo succession. Other areas, particularly those where *Miscanthus* is present, may be set back by fire and thus removed from consideration for swiddening for several more years. Some areas, because of other intrinsic site factors, such as depth of soil or previous good harvests, may be returned to and cleared on a more regular basis than other areas. It appears that at present there is no shortage of sites suitable for agriculture, and no apparent pressure on the length of fallow periods. Whether this is changing over time is difficult to judge.

F. Parameters of Disturbance: Scale

The scale on which swiddening occurs is relatively small, scattered plots averaging about .25 hectares. As demonstrated however, the scale of clearance varies with the type of vegetation on a site; large swiddens tend to be cleared out of early successional vegetation, while small ones tend to be cleared from primary forest, though they may be cleared from any other type of vegetation.

The Ayta practice of clearing one swidden adjacent to another may, as a factor of disturbance, increase the total scale. Such adjacent clearance of swiddens occurs in 23 percent of the sample (N=9). As Frake (1962:58) has observed:

> the practice of clearing large areas adjacent to previously cleared areas increases, under certain conditions, the probabilities of succession to grassland instead of forest, thus removing the land from future swidden cycles.[30]

On an aggregate scale, only an estimate can be made of the total area cleared for swiddens annually within the research area. In the sample of 39 swiddens, 22 were found to be newly cleared during the 1980 dry season, a total of 8.65 hectares. For estimation purposes I will assume 9 hectares since size of swiddens tended to be underestimated. Assuming this represents between 40-50 percent of the total within the research area, it can be estimated that from 18 to 22.5 hectares are cleared annually in an area of 68 sq. kilometers. This represents between .26 percent and .33 percent of the total area.

G. Parameters of Disturbance: Intensity

Mueller-Dombois and Ellenberg (1974:393) have stated that "The most complete disturbance leading to secondary successions is the conversion of forest to agricultural cropland." The intensity of disturbance of an area as a result of shifting cultivation is a primary determinant of the succession which will result. However, "the effects of cultivation on tropical forest ecosystems differ considerably depending on circumstances" (UNESCO 1978:467).

There are a large number of parameters which may determine the intensity of disturbance resulting from swiddening, and hence the nature of the fallow growth which follows. It has been noted (UNESCO 1978:218) that "The disturbance of the forest has three phases: felling of all or most of the trees on a patch of forest, burning, and cultivation of one or more crops." If we can ignore the specific reference to forest in the preceding, this statement provides for the three primary components of disturbance intensity in swidden agriculture.

---

30. In the Cabusilan area this depends on whether *Miscanthus* or *Imperata* becomes established.

In clearing a swidden, the cultivator attempts to remove most or all of the above-ground biomass. The degree to which this is accomplished depends on the sort of community which is cleared and the extent to which vegetation is actually removed; often there are plants which the swiddener need not expend the labor necessary to eliminate. In a swidden cleared from *Miscanthus* or *amukao* it may be possible to remove the entire stand. In swiddens which are cleared from early secondary forest dominated by fast growing softwood trees, the undergrowth is removed and the trees frequently just coppiced. They may then serve as supports for lima beans. Whether or not these small trees survive, and how densely spaced these survivors are, has an important impact on the succession to follow. Kowal (1966:399) observes that:

> If the living stumps . . . are closely spaced, a dense growth of stump sprouts develops rapidly, preventing the entrance of many new species. . . . Where the stumps are more widely spaced a rich temporary community develops.

Though Kowal speaks of stumps, the same applies to coppiced trees. In swiddens cleared from old secondary or primary forest, as well as those cleared from early secondary forest, many of the smaller trees are retained. Large dipterocarps or other trees may either be cut down or left untouched, depending in part on whether their crowns will cast too much shade.

Though the above-ground biomass of a plot may be removed completely, it requires noting that none of the below-ground biomass is affected. Thus a large number of plants and trees may survive initial clearance, particularly among light-tolerating or light-demanding plants. However, as the drying-out of the cleared swidden proceeds, a large percentage of the plants may not survive.

With burning, selective pressure is again exerted on the remaining plant community. The actual intensity of a swidden fire may vary greatly depending upon the type of material burned, its moisture content, wind and slope. *Miscanthus*, *amukao* , and young secondary forest all burn very differently. With *Miscanthus*, drying is generally very complete; the resulting fire is hot, but very rapid, and the dried leaves and stems are fully consumed. With *amukao*, drying is not as complete, particularly for the pseudostems; burning is rapid though subdued, and numerous soggy pseudostems remain. Swiddens cleared from woody growth, whether primary or secondary, generate the highest sustained temperatures and have the greatest impact on below-ground biomass.

The Ayta generally wait for light breezy days to begin burning their swiddens. Wind, of course, contributes to both the speed and intensity of a burn. On steep slopes strong updrafts may occur, further intensifying the burning process. The greater the intensity of burning, the deeper heat penetrates, and the greater the impact on sub-soil biomass. Also, seeds present in the soil are more likely to be killed, thus further limiting the recovery of a site. UNESCO (1978:220) states that:

> Many seeds, and most seedlings will be killed, and trees are damaged by fire, but organs located more than a few centimeters under the soil surface will not normally be affected. There are great differences in fire tolerance between species and fire is thus a selective factor, tending towards a simplification of the floristic composition and a complication of the successional pattern.

Elsewhere UNESCO (Ibid.:220) states that "Fire affects the regeneration (of a community) both directly through the burning of seeds, seedlings and trees, and indirectly, through its action on the soil." This edaphic effect is of primary importance: it too is a function of the intensity of a fire. Both the physical and chemical properties of the soil are altered by burning. Referring to the physical properties of soil, Budowski (1966, in UNESCO 1978:220) notes that "There is often a deterioration of structure due in part to the dismembering of soil aggregates resulting in clogging of the porous space." In terms of nutrient status, Farnworth and Golley (1973:150) state that "When a forest is burned, nutrient cations and phosphorous from the vegetation are accumulated in a thin layer of ash on the surface soil, resulting in a temporary increase in the fertility of the soil." He continues, noting (Ibid.:151) that "most nutrients are immediately made available for plants at the soil surface, but can be lost by leaching or erosion."

Leaching of nutrients from the soil inevitably follows the clearance and burning of a swidden plot. The rapid recycling of nutrients is one of the most oft cited features of tropical forest ecosystems. The decomposition following the cutting and burning of a swidden releases those nutrients which have been sequestered in the biomass of a plot. The result is ash. This of course is what the swidden farmer wants; it is this soluble ash layer which provides nutrients for crops. But it is their solubility which allows these same nutrients to be leached and lost from the system (see Bartlett 1956:705). This tendency is most marked in areas of high rainfall: Zambales is certainly such an area. Spurr and Barnes (1964:446) note that "In the higher rainfall areas, it may take only six months to leach a cleared forest soil to an unproductive state."

Patterns associated with cropping are also important factors in affecting the intensity of disturbance. These include (1) weeding, (2) the seasonal pattern of planting, (3) types of crops, (4) spacing of crops, and (5) length of cultivation.

Weeding acts as a selective agent against certain species of plants, not randomly but rather as the product of purposeful human action. The cycle of weeding has already been described; two major episodes, with several minor ones following as needed. Kellman (1969) found that repeated weeding on one plot in Mindanao had "a deleterious effect on the regrowth species population, each cleaning cycle serving to reduce the quantity of viable seed stored in the soil." In weeding, an effort is made to remove all parts of the plant, particularly the roots. The exceptions are *Miscanthus*, *amukao* and *Chromolaena*. All three are difficult if not impossible to remove without exerting enormous amounts of energy and completely disrupting the soil. *Chromolaena* has a deep, woody tap root which easily survives burning and is the source of vigorous new growth. The root structures of the other two are even more tenacious, and foster vigorous flushes of growth after burning. These are kept down during the cultivation period, particularly *Miscanthus* and *Chromolaena*, but they return rapidly once ignored. That these three are not subject to control by weeding, but can only be cut back, places them in a prominent position in the fallows which follow.

The seasonal pattern of planting is another primary determinant of disturbance intensity. Among Ayta, the clearing, burning and planting of swiddens takes place from the middle of the dry season into the rainy season, the usual pattern in monsoon tropical swidden systems. This pattern has major implications with regard to the intensity if disturbance. Post-burn swidden soil is soft, friable and exposed. This is the point at which the rains begin. Though a swidden may already have been planted, these plants have not yet become established and are thus not able to provide any cover for the soil. Crops such as corn, and to a lesser extent rice, never establish a full cover. The extreme rainfall in Zambales has already been noted, particularly in the months from June to September. Torrential rains are the norm rather than the exception, several consecutive days of such rain being normal during the passage of typhoons. The soil is therefor exposed to the full impact of these rains at the point when it is most vulnerable. When the slope factor is further considered, the erosional potential increases exponentially.

The types of crops planted is another important factor in determining the intensity of disturbance. The crop complex may act in two ways: physically and chemically. The physical aspect is seen in relation to the ground cover achieved by certain crops, which either protects or exposes the soil to erosion, compaction and insolation. Sweet potato, for instance, once established, provides a thick, complete insulating ground cover. *Xanthosoma* and taro provide cover to some degree, but are generally dispersed. Rice provides relatively little protection, though the meristems significantly impede surface flow. Corn is perhaps the least beneficial crop in providing ground cover, and cassava is little better.

The chemical aspect of disturbance intensity refers primarily to nutrients which certain crops either contribute to, or take from, the soil of a swidden plot. Brookfield (1968:418) citing Kostrowicki's (1965) crop typology, refers to the "exhaustive, extractive, soil-depleting crops." Grain crops such as rice and corn fall into this category. Harris (1972:403) has noted that seed crops "tend to make heavier demands on soil nutrients." Legumes such as pigeon pea and lima bean may, on the other hand, contribute to the chemical status of the soil by fixing nitrogen.

The spacing of crops is another important though frequently overlooked determinant of the intensity of disturbance. It has a direct effect on rainfall interception and on surface flow, the two primary components of soil erosion. Sweet potato cuttings are spaced about 1 meter apart. Corn is spaced an average of 1 to 1.5 meters apart, and

Xanthosoma .5 to 2 meters apart. Reference to Plates 33 and 34 illustrate this clearly. Only in the case of sweet potato is a full protective cover achieved, and this requires several months.

The length of period of cultivation before a swidden is allowed to revert to a natural successional pattern has been recognized as one of the most significant factors in determining the intensity of disturbance. In this regard Richards (1952:401-2) is worth quoting at length:

> If secondary succession is allowed to start immediately after the forest has been cleared, there is little opportunity for soil erosion and a closed cover of vegetation is formed in a few weeks; the process of soil impoverishment is arrested before it can go very far and soil and microclimate begin at once to return its to their original condition. If on the other hand there is a long period of cultivation after clearing, particularly if the crop does not provide an adequate ground-cover, the long exposure of the soil, together with the losses of plant nutrients in the harvested crops, leads to such large changes in the structure, humus-content and nutrient status of the soil that the time needed to restore the equilibrium between soil and vegetation, even when there is no large-scale erosion, becomes much greater. The importance of the length of the cultivation period and the extent of humus destruction to the course and length of the subsequent succession are thus easily understood.

A good example of this is Street's (1969:106) citation of Freeman's (1970) findings. Freeman:

> ascertained that cropping for two years in succession raised considerable hazards of erosion and displacement of forest by a highly combustible savanna of *Imperata cylindrica*, whereas cropping for but one year coupled with a fallow of 11-14 years markedly lessened these hazards.

In Zambales the length of the cropping cycle is generally only a year. Certain crops, particularly sweet potato, *Xanthosoma*, taro, and pigeon pea, may be harvested for longer periods. However, no additional clearance is required in the second year, only a small amount of weeding, and the effect is relatively benign. Most of the damaging effects in the Ayta swidden system occur in the first year of cropping.

There are three points, then, at which the vegetation of a plot, or disseminules, are subject to selective pressures: clearing, burning and cropping. Under the category of cropping are the effects of weeding, erosion, shading and competition. All these have a major influence on the resulting fallow.

All aspects of swidden disturbance intensity so far considered have one thing in common: they foster erosion. Erosion is a critical factor in determining the nature of a successional sequence, and in limiting the sustainability of particular swidden systems. Harlan (1972:239) notes that "The most critical feature of agriculture in the uplands of the wet tropics is the maintenance of soil fertility and structure."

A critical feature determining erosion is the degree to which the soil is protected from the direct impact of raindrops. Temple and Murray-Rust (1972:195) state that "Splash and sheet wash detaches, transports and redistributes top soil downslope. It generally operates selectively carrying away first the finer particles and lighter organic matter most valuable to plants." Tamhane et al. (1959:231) observe that:

> Soil detachment, a phenomenon responsible for soil erosion by rainfall, is related directly with the intensity of rainfall as the enormous kinetic energy of the falling rain drops is dissipated into the disintegration of soil particles which are subsequently transported and lost in runoff water as soil loss.

Soil covered either by leaf litter or by a canopy, whether high or low, is protected from this direct kinetic energy. Butzer (1974:73) cites Hudson's (1971) estimate of "the erosive forces of tropical rains to be 16 times greater than that of mid-latitude rains." Temple (1972:204) cites the findings of Hudson and Jackson (1959) that, "in conditions of subtropical thunderstorms with high intensities and large drop sizes very severe erosion occurs when there is no protective cover." As we have seen, many crops planted by Ayta do not provide adequate soil protection. This, added to the friable structure of the soil following burning, fosters a situation of extreme erosive potential.

The slope factor multiplies the potential for erosion even further. Butzer (1974:59) notes that "Steep slopes favor runoff over infiltration and accelerate overland flow, particularly on long slopes that increase water velocity as a function of time and mass. . . . (Horton:1945)" Kowal (1966:398) found in the Cordillera of northern Luzon that:

> On the steep slopes . . . *kaingin* [swidden agriculture] resulted in severe erosion, which in combination with the more intensive fires caused by updrafts, effectively prevented the regrowth of montane forest.

The situation in Upland Zambales--intense rainfall, steep slopes and exposed, friable soil during the wettest months--presents the ideal setting for extreme rates of sheet erosion and soil loss.

CHAPTER 7

BURNING

A. The Ecology of Fire in Upland Zambales

The degree of uncontrolled burning which occurs in upland Zambales is remarkable. It is a primary factor in maintaining so much of the vegetation in the area at earlier stages of succession. In this chapter all references to burning will concern the random burning of grassland and other secondary associations. Swidden burning has already been discussed and, I suggest, differs considerably in terms of its ecological significance: all the parameters of disturbance--scale, frequency and intensity--differ markedly.

Farnworth and Golley (1973:120) note that:

> Fire is among the most important of the physical factors influencing succession and, although we have a large body of descriptive information on the effects of fire in certain communities, we lack information about its effects on the successional process.
>
> Factors such as intensity of fire, temperature, frequency of burning, soil temperature during the fire, and effects of high temperature on propagules and on physical and biotic properties of the soil can be critical in determining the direction of the succession.

The extent of grasslands and other early successional associations in upland Zambales has been described. It is estimated that 35-40 percent of the total research area is covered by grassland associations. The maintenance of these communities is dependent on, and a function of, fire. Fox (1953:183) has stated that "Undoubtedly, the whole region was once heavily forested, but due to the continual burning by the Negritos, the tough weeds and grasses . . . have invaded the plateaus, and have become dominant." During the long dry season these grasslands become very dry, at which point they burn easily. Approximately 35 percent of the early successional vegetation, or 18 percent of the total research area, were subjected to burning during 1980. These figures do not include numerous small fires which occurred but, for logistical reasons, were not fully documented.

Fires within the research area occur primarily in grassland communities. This includes both *Imperata* and *Miscanthus* associations. On occasion other types of associations are also affected. Running fires often burn into mixed stands of *amukao* and *Miscanthus*, as was witnessed on several occasions. Pure stands of *amukao* seem able to resist burning very well. Only rarely are areas of secondary forest affected by burning. Primary or late secondary forest appear never to burn.

Fires may be started for a number of reasons. Formerly, when pig and deer still populated the Cabusilan area, fires were set to drive game. References to this are common in the literature. Garvan (1963:69), describing group hunting in Zambales, states that, "Whenever it can be done, the grass is set on fire in different places." Taylor (1936:222), describing a hunt in the early 1930s which was almost certainly within the present research area, gives the following somewhat fanciful account:

> A mile to the right a column of smoke began to curl up. The beaters had fired the cogon [*Imperata*] grass. In a minute or two the flames reached a thicket of bamboo and the hollow stems began to burst, slowly at first, then rapidly like musketry. As the wind whipped the flames deeper into the thicket, the explosions came like a rolling blast of shotgun fire.

Such drives still occur on occasion, but are very rare. One was observed from a distance taking place on the upper southwest slopes of Mt. Pinatubo. More commonly today an area may be torched to drive out some small animal such as a snake or lizard. This was observed on a number of occasions.

Perhaps the most common reason Ayta set fires is to make walking through high grass easier. *Imperata* generally occurs in thick stands, which can only be moved through with great discomfort. Its finely serrated leaves cause cuts and itching. Thus, Ayta intending to walk through such an area may burn it first. I witnessed a dramatic example of this while hiking along the margin of the grassy Kapisan area on March 24, 1980. Passage through this area was extremely difficult. Thus, as we walked along the ridge, a series of seven fires were set. The wind was strong and extremely gusty. In less than ten minutes over four hectares had been cleared away. The hike back was much easier.

It was often difficult to get clear responses from people as to why they would set fires. One reason is that fires may be set for no reason at all, or just for the "fun" of it. One informant stated that "We don't do it for fun exactly, we just like to see a place burn."[31]

A large number of fires are also set accidentally. One case was observed where a fire set to smoke out honey-bees escaped, burning up a large slope before extinguishing itself. When travelling, Ayta often carry lighted faggots for starting campfires or for smoking. Both embers and cigar butts are frequently discarded along trails, often while still smoldering.

In Chapter 5 evidence for the antiquity of deforestation in this area was presented. Presumably much of this deforestation resulted in the conversion of forest to grassland, as continues today. By 1907 large portions of the research area had become dominated by grasslands, as photographs taken at that time demonstrate. Burning was clearly occurring at this time. We can recall Smith's statement (1909:23) that "The coarse, rocky soil at the base of Mt. Pinatubo contains practically no vegetation, the very small amount of grass which starts being periodically burned off by the Negritos." He also refers to "The bare, burned surface of the mountain" (Ibid.:23). In Taylor's 1936 description of a hunting trip in an area which sounds like the head of Maagə'agə' canyon, he noted that "A mile above us was the knife edged summit of a ridge that had been recently burned over" (1936:220). Aerial photographs taken in 1946 also show large areas having been burned.

---

31. Paraphrased translation by Mrs. Aloma Tima.

There are a number of features of grasslands which contribute to their pyrophilous character. Vogl (1977:152) has noted the "decay resistance of grasses." He states (Ibid.:151) that "grassland plant debris often accumulates faster than it decomposes. . . . [This is] added to the litter layer at the end of each growing season rather than becoming a functioning increment--as in woody vegetation types." Brown noted (1919:18) that "Dead leaves accumulate quickly in stands of *Imperata* or *Saccharum*, so that these grasses become very inflammable during the dry season. At such times a grass area is usually burned over completely." Also significant are those features which allow certain species of grass to survive burning, the most important being the system of dense underground rhizomes in genera such as *Imperata* and *Miscanthus*.

B. Parameters of Disturbance: Frequency

Chapin and Van Cleve (n.d.:1) note that "Fire frequency places definite constraints on the types of growth forms that are successful in a given environment." Further, "when fire frequency is either increased or decreased as a result of man's activities, there is usually a rapid change in growth form dominance (e.g., Daubenmire 1968; Bragge and Hulbert 1976)" (Ibid.:3). The parameter of frequency of burning is intimately related to the entire functioning of most grassland ecosystems. In Zambales, the frequency of burning is geared closely to the seasonal cycle of wet and dry months. During the dry months grasslands become tinder-dry and ignite easily. This is when burning occurs, in particular between the months of January and April. A large percentage of grassland communities within the research area are burned every year, at least 50 percent. Signs of past burning are everywhere in abundance. Aerial photographs taken in 1946, and again in 1967, both show large areas of burned-over grassland. Exposed ridges appear to be more susceptible to burning than protected slopes. This is due to differences in species composition, higher wind conditions on the ridges, and a dryer edaphic environment. Wind increases evapotranspiration, further drying the grasses, and serves to drive the fires which do start. Vogl (1977:150) observes that:

> winds . . . tend to dry the grassland fuels. Grasslands occurring on light soil types with poor water-retaining capacities are particularly susceptible to drying by transpiration and evaporation. Drying is often aided by sunlight penetration, which can reach ground levels as the winds bend, whip and separate the plant cover.

C. Parameters of Disturbance: Scale

Considerable variability exists in the scale of fires which occur within the research area. They may be localized or may cover large areas. A fire observed on May 6, 1980 on the upper slopes of Mt. Pinatubo burned approximately two square kilometers. Another fire which occurred in mid-January on the Lomboy plain burned an area of equal size. Still another large fire on the slopes above Maagə'agə' burned approximately one square kilometer. Fires of from one to ten hectares are most common. The deliberately set fire observed on Kapisan on March 24, 1980 burned about four hectares. Map 6 indicates areas subjected to burning during the period of research. This represents a substantial underestimate; many areas burned previous to the research period and supporting regrowth were not included, as their boundaries were difficult to distinguish.

A number of factors control the extent of individual fires. Within the research area topography is a factor of major significance. This area, except for the Lomboy plain and the higher mountains, is little more than a series of sharp ridges and steep valleys. Such terrain may impede the spread of fires, as has been noted by Vogl (1977:269). Fires spread quickly upslope, but slowly down. Daubenmire (1968:214) has stated that:

> fires move faster upslope than on level ground, advancing still more slowly when moving downslope. Fire spreads up a 10° slope twice as fast as on the level, and progresses four times as fast up a 20° slope. Where the slope is very steep, the rate of spread may exceed 65 Km/h.

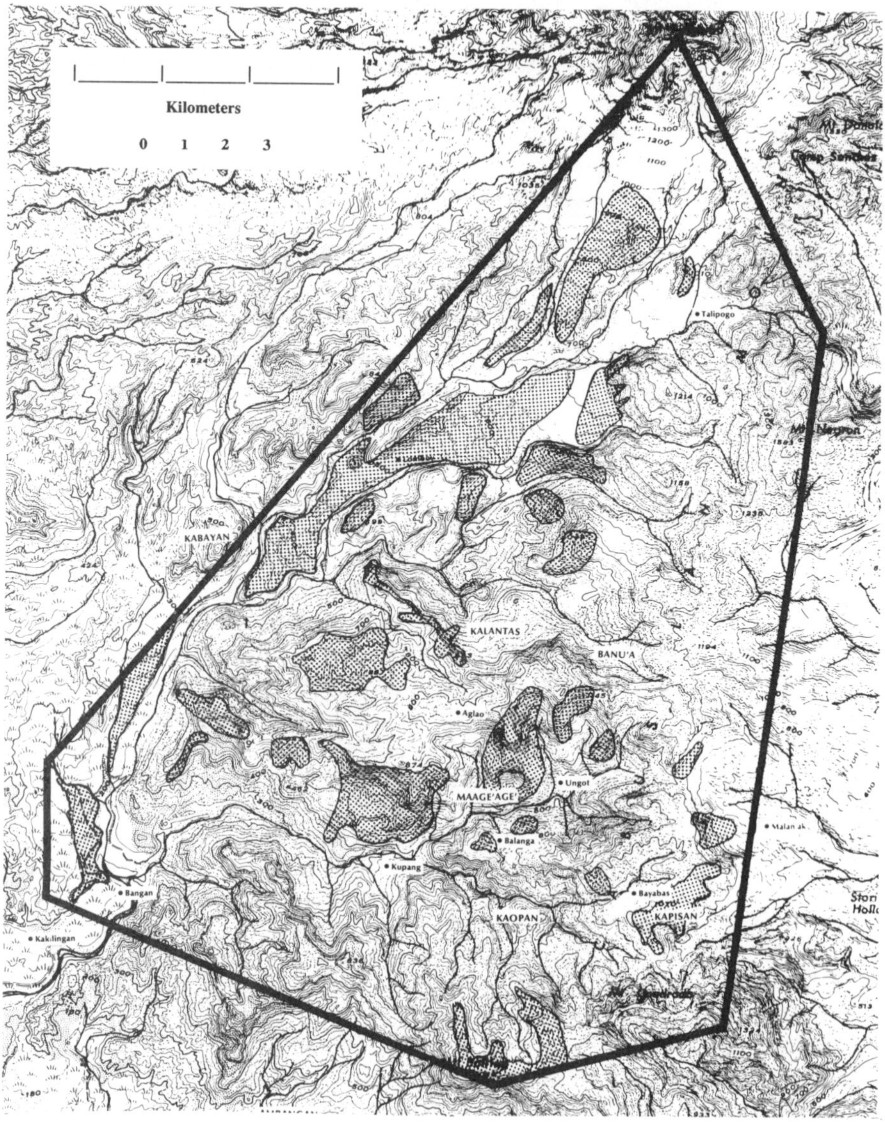

Map 6. Approximate Delineation of Areas Burned--1980.

A frequently observed pattern within the research area is for a fire to begin in a valley, move up the slope, and spread along the ridges.

Wind is a factor which affects not only the frequency of burning, as noted, but also the scale. On windy days fires may move at tremendous speeds, covering large areas in a very short time, as was seen during the March 24 Kapisan fire. Daubenmire (1968:216) notes that "The rate of fire spread varies approximately with the square of wind velocity." As in swidden fires, updrafts and convection currents may also act to accelerate burning.

The extent, composition and position of vegetational communities may also be an important factor in determining the size of an area burned. Large unbroken patches of *Imperata*, such as are characteristic of Lomboy and Kapisan, may burn almost entirely in a single fire. In rough, broken areas, with a mosaic of vegetation communities, fires are more contained. Here can be seen the significance of historical changes in the distribution of different vegetational associations. Speaking of the Cordillera Central in northern Luzon, Kowal (1966:400) notes that "Perhaps the most important factor making . . . frequent fires possible was the gradual removal of the montane forest as a continuous matrix insulating the patches of *kaingin* from broadly extending fires." Frequently, on protected slopes, communities of *amukao*, secondary forest or *Miscanthus* are present. Depending on conditions, these may not easily burn. If a fire is burning down a slope it may simply burn itself out on encountering a less combustible association in moister edaphic conditions. It is noteworthy that *Miscanthus* is less susceptible to burning than *Imperata*. It grows in dense stands on protected slopes and its thicker stems make it more resistant to desiccation.

D. Parameters of Disturbance: Intensity

The intensity of disturbance resulting from the burning of grassland is much less than in communities subject to other types of disturbance. This is in part due to the "structure" of these fires. Daubenmire (1968:213) notes that "Grass fires are characterized by a rather narrow zone of flames advancing across a finely divided rather homogeneously dispersed fuel." Vogl (1977:153) states that:

> Dry grassland fuels tend to be flashy, igniting readily and burning quickly. In addition, grassland fires generally produce a narrow belt of flames and pass rapidly because of the low growth, open terrain, level to gently rolling topography, and the general presence of winds. As a result fires seldom tarry long enough to build high surface temperatures. The highest temperatures are usually produced well above the ground at the apex of the flames and above, and are rapidly dissipated by winds. Grassland fuels are normally converted to ash so rapidly that one can follow immediately behind most fires without experiencing any discomfort from heat, smoke, prolonged burning or hot ashes, and the soils are generally cool to the touch.

There is an important relationship between the frequency and intensity of grassland fires: simply stated, the greater the frequency, the lesser the intensity (e.g., Vogl 1977). This relationship, resulting from the accumulation of fuels over time, has received a great deal of attention in the fire ecology literature. San Jose and Medina (1975:262), speaking of Philippine grasslands, note that "long protection increases the amount of litter accumulated. Subsequent fires are much more destructive." Daubenmire (1968:215) sees "The kind, amount and disposition of fuel that has accumulated since the last fire" as one factor determining the "amount and rate of heat released as vegetation burns" (Ibid.:214). In Zambales, with burning occurring every year or two, intensities are generally low.

Other factors affect the intensity of burning as well. Wind, mentioned as a factor determining both frequency and scale, also affects intensity. This has been noted by Daubenmire (Ibid.:214) and by Vogl (1977:150), who notes that "The low growth forms of the plants contribute to the open nature of grasslands, and with the large areas of continuous vegetation, allow for the unrestricted flow of winds." Grasslands "are windy places . . . and the winds not only provide oxygen for grass fires but drive headfires at alarming speeds, even with heavy fuel moisture" (Ibid.:150). Within the research area, the high ridges and valleys and the Lomboy plain are very windy places.

The amount of moisture present in vegetation and in the soil is another factor determining the intensity of disturbance due to burning. Quite obviously, dry fuels burn quicker, hotter and more completely than those with a higher moisture content. During the long and severe dry season in Zambales, lack of rains, combined with wind, bring about intense drying. By January the slopes appear parched and brown--optimal conditions for burning. There is considerable micro-environmental variability in the moisture retention potential of different areas. Protected slopes, especially those with deep soils, receive neither as much sunlight nor as much wind. These areas generally support *Miscanthus*, *amukao*, or other secondary associations rather than *Imperata*, and appear to retain moisture more effectively than the latter.

The length of disturbance resulting from the burning of grassland is a very distinctive feature of this type of perturbation. There are two aspects of this: (1) length of initial disturbance and (2) post-fire site changes. As noted, grassland fires burn with extreme rapidity. Once established they are characterized by a rapidly moving but narrow band of flame. Fuels are immediately volatilized and reduced to ash at which point, as noted, it is possible to walk through an area. Daubenmire notes (Ibid.:218) that "the duration of maximal values is so brief that direct influence of grass fires on soil would appear negligible." The actual burning process, then, is extremely brief. Soil physical properties are not greatly affected by grassland burning. That is, fires are not of sufficient intensity to cause changes in soil structure, such as would result from a swidden fire. Heat does not penetrate deeply, rarely more than a few centimeters. Daubenmire (1968:256) states that "From the surface downward, the temperature gradient is so steep that direct alteration of soil structure or chemistry by fire is inconsequential in grasslands. Kucera (n.d.:15) states that "Studies show that soil temperatures 2.5 cm below the surface rise very little in the active fire zone (Scotter 1970)." This is particularly true for soils with low moisture content, which is the case in Zambales during the period of burning. Daubenmire (1968:217) notes that "The drier the soil, the higher its surface temperature when grass burns, but at the same time the low moisture content reduces the downward conduction of heat."[32]

It is after burning that the most protracted disturbance occurs. The elevated temperatures following burning are not sufficient to affect soil chemistry and structure, but do have a major impact on soil biota. This is a consequence not of the fire itself, but of the higher sustained soil temperatures which result from increased insolation following burning. Recently burned areas are covered by a thin, uniform layer of dark ash. Komarek (1967:146) notes that "This . . . absorbs more heat from the sun." Vogl (1977:156) observes that:

> The blackened soil surface is known to absorb solar heat, thereby causing higher temperatures than are produced on comparable unburned soils (Daubenmire 1968). Some post-burn soil surfaces may actually reach higher temperatures from direct insolation on the exposed and darkened surface than they did during the fire.

In the context of upland Zambales, darkened, exposed soils are subjected to insolation for long periods during the clearest and warmest part of the year, raising soil temperatures considerably. However, within a few days of burning a new flush of growth occurs, fostered by moisture retained in roots and stems. Within a few weeks light cover is established. Vigorous growth resulting in reestablishment of full cover commences with the first rains.

Erosion does not appear to be a significant component of disturbance following burning, particularly in grasslands. When such an area has been burned the rhizomatous underground root masses of *Imperata* and *Miscanthus* remain intact. Protruding charred meristems and new flush growth impede surface flow. Vogl (1977:160), negating the incidence of soil erosion following grassland burning, observes that "the remaining basal crowns, fibrous and

32. This is contrary to the claim by Zinke et al. (1978:144) that in swidden burning, soil moisture insulates. Also Komarek (1967:146) states that "soil, particularly damp soil, is a very good insulator and only in unusually severe or intense fires where a *great amount of fuel is burned* is the soil heated appreciably" (emphasis original).

extensive root systems, ash, charcoal and litter usually protect the soils from severe and massive erosion." Rains at the end of the dry season begin gradually, fostering vigorous growth before the most intense rains occur. Thus the kinetic energy of falling raindrops is also impeded. This is not to imply that soil erosion does not occur; it must to some extent. Relative to other forms of disturbance however, it would appear slight.

The impact of burning on the physical and biotic properties of grasslands is very different from that following disturbance in other types of associations. Generally speaking, the impact is more a function of the frequency of disturbance than of intensity or scale. Komarek (1967:148) states that "recurring fire acts as a natural selective agent." Daubenmire (1968:248) observes simply that "Species growing together may respond very different to the same fire." Grassland burning has a direct effect on biotic components in selecting either for or against certain species. The point has been made repeatedly that graminous species such as *Imperata* and *Miscanthus* are adapted to a pyrophilous regimen, and that repeated burning eliminates other species, with the exception of a few legumes and ferns.

Fires within the research area do not generally occur in primary or secondary forest growth, but are limited mainly to grassland and *amukao* vegetational associations. Thus, it cannot be said that grassland burning has actually been a cause of deforestation. Rather, it has served as an instrument of maintenance, preventing the return of areas so affected to later stages of succession. Repeated burning results in a continuous process of biotic sterilization wherein non-fire-tolerant species are selected against.

CHAPTER 8

LOGGING

A. Logging in Upland Zambales

Logging has been a factor of disturbance in the Cabusilan area for a very long time. Where it has occurred the impact has been great. In attempting to explain the historical sequence of environmental degradation within the research area, logging cannot be omitted: it has had a significant impact on the Ayta resource base, and on the evolution of the contemporary agroecosystem.

Two types of logging have occurred in the Cabusilan area, following an historical sequence. Before World War II all felling operations were non-mechanized, done primarily with the aid of carabaos. After World War II mechanization occurred and larger-scale operations became possible. The parameters of disturbance vary greatly for these two types of logging. In the following discussion I will review this sequence and consider the relative impact of each with respect to factors of disturbance.[33]

The history of logging in Zambales may date back to the pre-Hispanic period, when valuable woods may have been traded with the Chinese, who were actively trading along this coast. Such trade would

---

33. Logging was no longer occurring within the research area during the fieldwork period, and thus was not directly observed. Much of the following description is based on interviews with Antonio Rodriguez, who directed the most recent logging operations and whose family has had the timber concession in the area since the turn of the century. Napoleon Vergara of the East-West Center, Environment and Policy Institute, also provided much valuable information on technical aspects of logging in this area.

have had a very minor impact on the forest resources of the area as only the most valued trees would have been very selectively felled. With the consolidation of Spanish control over the area in the early seventeenth century, historical records suggest that the lumber trade was well institutionalized. We can recall the statement by Perez in 1680 (Blair and Robertson 1909, 47:292) concerning the earnestness with which the local population applied themselves to cutting timber near Bagac on the west coast of present-day Bataan (see p. 58). Zuniga, remarking in 1803 (Zuniga 1973:346) on the lumber trade in Pampanga, stated that "the mountains . . . provide them . . . with much costly lumber not only for building good houses but also to furnish attractive items in their trade with Manila." Later he quotes from the Franciscan "Historia," in reference to Zambales. Noting the "Moro" threat along this coast it states: "These enemies of peace . . . cause great damage to the Zambals because they could not transport the hard woods in rafts nor could they bring their merchandise in small vessels to Manila" (Ibid.:377). Other writers note that the early Sambal paid their tributes to the Spanish with lumber. During this early period it appears that no lumber was being taken from within the Cabusilan research area, with one possible exception. It is conceivable, as noted in Chapter 5, that the Lomboy plain and the area near present-day Kakilingan may have supported stands of molave which, due to ease of access, may have been depleted very early.

The quality of lumber in Zambales is very high, being dense and highly durable. This is frequently noted in the historical record. The *Report of the Philippine Commission* (War Office 1908:475) states that "The province is splendid in forestal products, and first-group timber,[34] such as molave, yacal, narra, acle, etc. abounds." Reed (1904:24) states that "The forests are especially valuable and besides fine timbers for constructional purposes they supply large quantities of pitch, resin, bejuco and beeswax." Most of the cutting done during this period was presumably very selective, only the most valuable species being taken.

Around the turn of the century, wood cutters from Aglao, a small Ilokano barrio at the base of Mt. Balitog, began cutting lumber

---

34. "First-group" refers to the four-level timber classification system used in the Philippines, based on quality and value. First-group woods are of the best quality.

on the lower slopes of this mountain, directly south of the research area, and in the area near the present Ayta village of Kakilingan. This logging was mostly carried out under the control of the Rodriguez family. According to Antonio Rodriguez, the present patriarch of the family, all logging at that time was "carabao logging." Trees were felled by hand, sawn into large beams on site, and dragged out individually with carabaos. Corroborating published sources from this period, he states that only special species were taken: nara (*Pterocarpus spp.*), tindalo (*Pahudia rhomboidea*), quijo (*Shorea guiso*), akle (*Albizzia acle*), apitong (*Dipterocarpus spp.*) and malaruhat (*Eugenia sp.*). Parker (1964:118) gives a very good description of this type of logging as carried out on the eastern, Pampanga watershed in 1908, a system very similar to that on the western side:

> Quite an extensive trade is carried on in lumber by the Filipinos of Floridablanca.
>
> The work of cutting the trees, shaping them, and drawing them to the valley with carabaos, is practically all done by Filipinos. Very little help is given them by the Negritos.
>
> The Negritos claim the trees and charge that the Filipinos take the timbers without paying the owners and often without permission or promise of pay.
>
> The persons who do the actual work of getting out the timber are only the peons of a few well-to-do Filipinos and Spaniards of Florida[blanca], who probably have authority for cutting timber.

One Ayta informant from the Mt. Balitog area, where this early logging occurred, complained that "There was much more forest here before. The Spanish cut it, then Rodriguez. There was no *amukao* before."

It was after World War II that profound changes took place. Two factors were of particular significance: (1) technological change, and (2) the lumber market.

In 1945, with the end of World War II, surplus trucks and bulldozers suddenly became available. Lumber concessionaires were quick to seize this opportunity and procured large numbers of them. This completely changed the complexion of the local logging industry.

Roads were pushed through into the foothills, beyond Kakilingan and south into other areas. Logging at this time was still relatively selective and the market for Zambales lumber was largely local. Very little was being exported to other parts of the Philippines. The so-called first-group high-quality trees were still most in demand, though a number of relatively high quality second-group trees were also now being exploited. There was not yet a market for trees of lesser quality.

It was after 1960 that logging activities became more extensive, due primarily to a change in local market conditions: third- and fourth-group logs were now being accepted. Logging was much less selective and thus more complete. Lower elevation areas previously logged were re-logged. It was at this point that roads were pushed up the canyons and ridges and into the research area. Aerial photographs taken in 1967 show logging roads covering the lower slopes of Mt. Balitog, into Maagə'agə' and over the top of Balitog. A road also went through to Lomboy and Kalantas (see Map 7). All this logging activity ceased shortly after the imposition of Martial Law in late 1972.

B. Parameters of Disturbance from Carabao Logging

The impact of the two forms of logging employed within the research area differed greatly. The parameters of disturbance characteristic of each--frequency, scale and intensity--were very dissimilar.

Carabao logging occurred in a setting of either undisturbed or culled primary forest. When talking about logging, it is difficult to speak in terms of frequency since it generally occurs in areas of primary forest. In the present case however, it is likely that a very gradual culling of high quality trees was occurring over a period of several hundred years in lower elevation areas of Zambales and, since about the turn of the century, around the margins of the research area. Recovery from any on-site damage due to logging operations was thus possible.

With carabao logging, the scale of operations, and thus of disturbance, was very minor. With the use of carabao, roads were not necessary to gain access to logging operations. Logging parties could move into an area on foot, thus avoiding additional clearance. The only damage would be that from skid tracks when logs were removed. This would certainly have upset the humic layer of soil, particularly if skidding trails were utilized repeatedly. Clearly though, such access routes were of very small scale. Severo (1949:371) notes that "carabao logging trails occupied a minimum of space in the forest, thereby causing a minimum of destruction."

It is difficult to assess the scale of disturbance at actual logging sites resulting from carabao logging. Most likely, dispersed gaps, analogous (though more closely spaced) to gaps resulting from natural wind-throw, were the characteristic pattern. The dispersed pattern of tropical tree species has been widely documented in the literature (Janzen 1970; Richards 1952). This would serve to maintain a dispersed pattern of logging, since only a few species were being taken. Severo (1949:371) describes the system as follows:

> A very limited amount, consisting of carefully selected trees, is cut each year. Since cutting is very light, it does not appreciably affect forest conditions of the area. Very limited openings are made in the canopy so as to promote the rapid growth of the lower story trees.

UNESCO (1978:219) states that:

> Selective logging, limited to one or a few species and to a minimum diameter, may produce relatively small and sometimes only a few gaps per hectare. Such logging will deplete the forest of certain species and dimensions of trees, but ecologically it will not constitute a major disturbance of the system.

The intensity of disturbance of carabao logging is likewise minor. As noted, roads were not necessary. Skid tracks may have served to channelize run-off and generate localized rill-erosion, but on a limited scale. And again, at those sites where trees were being felled, the effect was analogous to windthrow. Large trees, in being felled, damaged other trees, as well as seedlings. The canopy was opened and the understory exposed to insolation. This fosters a flush of growth of young suppressed forest trees and secondary species. Spurr and Barnes (1964:441) observe that:

> Logging is similar to windthrow in that the cutting of commercially valuable trees tends to remove the overstory and release the understory. Many understory hardwoods cut in brushing operations connected with logging will resprout to form vigorous and fast-growing stems competing for overstory space in the developing new stand.

Slatyer (1976:13) states that:

> disturbance is slight over a small area . . . when an area of forest is harvested by traditional single tree logging methods. Light and soil water nutrients are only slightly increased over a small area and few individuals are killed by the disturbance. The gap is filled usually by vegetative growth of the surrounding adults or by replacement of the dead adult by a growth of offspring of late successional species that are already present as suppressed individuals. Few early successional species invade successfully because the area is small and the level of resources is low.

Reyes (1959:44) cites observations made in 1923 on the western slopes of Bataan, in an area subjected to carabao logging:

> The forest on the western side of Mt. Mariveles is very much drier than that on the eastern side. On the lower part of the area where animal logging was employed, patches of green timber were left standing and satisfactory reproduction was taking place, especially of apitong and white lauan.

With this type of logging, the area surrounding a felled tree is one of intense activity, as logging crews cut and shape the logs before removal. Much of the surrounding undergrowth is disrupted in the process. This activity is, however, of very short duration, and the area is soon allowed to begin the succession process.

Where logging activity is excessive, even carabao logging has destructive potential. Brown and Mathews (1914) present one such example in their description of the impact of logging in eastern Bataan Province. This was before the mechanization of logging, and the damage reported can only be due to carabao logging. Circumstances here differed from those adjacent to the research area however. This area borders Manila Bay and access to the Manila lumber market was much easier. Logging was apparently less selective, and a much higher percentage of dipterocarps was apparently taken. Further, logging had been occurring here on a larger scale for a much longer time. We can recall that the Malong revolt of 1660 began as a strike by 1000 wood-cutters working in the Pampanga-Bataan area. Brown and Mathews (Ibid.:536) describe the situation thus:

> the main canopy of the forest has been almost entirely removed, only 4 dipterocarps above 50 centimeters in diameter remain on the hectare and . . . the bulk of the stand numerically lies in the tree classes from 5 to 40 centimeters. . . . The sudden exposure to full sunlight results in the death of a great majority of the trees within a very short period.

They provide an example (Ibid.:536) of one 250 square meter plot before and after cutting. Before logging there were 1130 tree seedlings, 259 of which were dipterocarps. After cutting there were only 191 seedlings left, 38 being dipterocarps. Mortality will increase due to insolation and other site factors as time goes by. They note (Ibid.:456) that:

> The cutting of the large trees results . . . in the breaking and killing of a large proportion of the smaller ones. The ground is opened up to such an extent that almost all seedlings are killed by insolation, while most of the small trees become unhealthy and finally die. All of the large defective trees which have been left have been killed by brush fires, except in the very recent cutting areas.

Clearly the potential for destruction did exist with carabao logging. It appears that the significant element in the Bataan case observed by Brown and Mathews was the greater market for lumber which fostered clearance on a larger scale. Logging on the western side of the mountains adjacent to the research area was not as destructive. Felling operations seem to have been more specifically targeted at particular high-value timber species. While the zones surrounding such operations were exposed to certain stresses--mechanical damage from felling large trees, insolation and short-term disturbance of undergrowth related to log preparation activities--the effect was analogous to that of wind-throw damage. The impact of carabao logging is to cull the forest of particular species and to create a dispersed patchwork of secondary communities. Nevertheless, ecosystemic relations within the forest would not be impeded in any serious way.

C. Parameters of Disturbance from Mechanized Logging

The impact of logging increased considerably following post-World War II changes in technology and in the lumber market. This impact was a function of changes in the parameters of disturbance.

As noted, surplus trucks and bulldozers became available following the war. These were small 2½ ton capacity trucks, fitted with winches on the front. This allowed the development of the so-called "Bataan-Zambales Logging System," of which the mobility of operations is a central characteristic. Also, this system evolved for operating in very rugged areas. Trucks can operate in level areas of forest without roads or, in steep areas, with only minor bulldozer cutting. Trees are felled downslope from the cuts so that they fall in an uphill direction, lessening their impact and reducing breakage. These are sawn into lengths of four or five meters and winched uphill to be loaded, using a system of pulleys.

The frequency parameter of disturbance varied with the area and type of forest being cut. As the market for lesser quality woods developed, lower elevation culled areas at the south and west margins of the research area were re-logged. As roads moved onto higher elevation areas, uncut primary forests were encountered. Some

19. Demonstrating use of medicinal plant—Kapisan.

20. Woman with child, collecting young taro leaves for a meal—Bayabas.

21. Father and son on fishing trip. Man holding *asido*—Kalantas.

22. Sharing honey with children—Banu'a.

23. Man with wild chicken traps—Maagə'agə'.

24. *Mamatiktik* dam-building activity. *Amukao* leaves being placed under rocks to impede water flow—Maagə'agə'.

25. Infant and grandfather—Bayabas.

26. Arrows (L. to R.): *balangat*, used for hunting birds and bats; *hawel*, for civet cat; and *dumpil*, for "defense." Photo by Navin K. Rai.

27. Woman shelling pigeon peas—Maagə'agə'.

28. Tying up rattan vines for lima beans to climb.

29. Swidden in Kupang being cleared from *Miscanthus* association.

30. Removing sprouting *Miscanthus* from recently cleared swidden—Kupang.

31. Recently burned swidden shortly before planting. Note clumps of sprouting *Miscanthus* and *amukao*.

32. Drying swidden, cleared from *amukao*.

33. Newly planted swidden. Note spacing of sweet potato cuttings and exposure of friable soil—Bayabas.

34. Recently planted swidden with corn, sweet potato, *Xanthosoma*, and *Dioscorea alata*. Note spacing of crops.

35. Recently planted swidden with corn, *Xanthosoma*, and sweet potato. Note inclination, spacing of crops, and regrowth of coppiced trees—Bayabas.

36. Kapisan; fire set on March 24, 1980. Less than two minutes after being set. Photo by Navin K. Rai.

37. Site of March 24 Kapisan fire, several weeks after burning.

38. Site of March 24 Kapisan fire, immediately after passage.

39. *Amukao* area in foreground, logged in early 1960s and subsequently swiddened—Maagə'agə'.

40. Summit area of Mt. Balitog, extensively logged during mid-1960s. *Imperata* dominant on ridges, *Miscanthus* on slopes.

scattered emergents in grassland and secondary forest growth were also cut, but this was a minor component.

The scale of disturbance was much greater following mechanization. Two aspects of scale must be considered: roads and logging sites.

Two types of roads are necessary in mechanized logging operations: major roads and minor, unmaintained access roads leading to and around logging zones. Two major roads led into the research area; one up the canyon toward Maagə'agə', up to Mt. Balitog, and to the western base of Mt. Lingay. The other went around through Lomboy and up into Kalantas (see Map 7). These roads were used continually, as long as logging activities in the area continued. Radiating for short distances out from these were access roads, generally no more than rough bulldozer cuts, abandoned once a particular area was logged over.

The scale of areas logged increased with mechanization and an expanded lumber market. No longer were single, dispersed trees being removed. Clear-cutting, at least of trees above a certain diameter limit, was now the rule. This caused the removal of entire groves except for young residual trees and inferior species. The size of these groves varied considerably. The important fact is that now large open gaps were being created.

In terms of total scale, 20 to 30 percent of the research area has been subjected to logging. It is not possible to precisely delineate where the boundaries were. Road cuts are still visible, and the ridges and valleys subjected to cutting are well known to Ayta. Informants can point out specific sites where stands of timber existed. But in the absence of detailed mapping we can only estimate the total.

It is in the intensity of disturbance resulting from mechanized logging where the most profound changes have occurred. As with scale, there are two aspects of intensity: disturbance by roads and disturbance at logging sites.

The primary access roads into the present research area constituted a disturbance of major intensity. As most of these were on slopes of moderate to steep inclination, the soil profile was laid open, the upper and lower horizons simply being removed and dumped downslope of the cut. This was followed by the repeated passage of loaded trucks and other heavy equipment, over a period of several

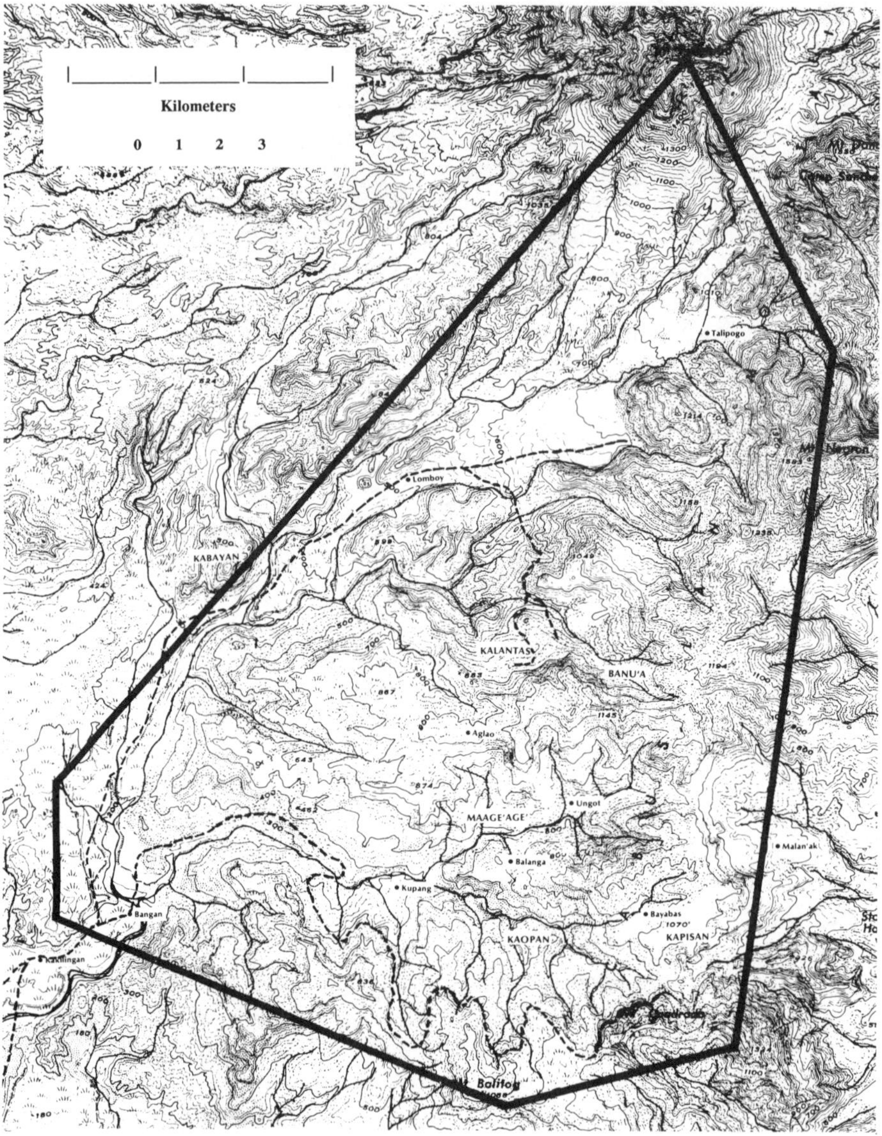

Map 7. Major Logging Roads

years. The chief danger of roads is that they foster erosion (Agpaoa 1975:410). UNESCO (1978:488) states that:

> In the Philippines construction of forest road beds after the wet season caused soil losses. Soil placed on the lower side of a cut, or left as fillings, is subject to erosion during the wet season. Erosion is particularly severe on steep slopes or where they have not been surfaced.

The intensity of disturbance on access roads is extreme, but relative to other types of disturbance it is focused and small-scale.

It is on-site disturbance which is most significant in mechanized logging, due both to severity and scale. Once a stand of timber has been located and deemed suitable for cutting, four stages of disturbance can be identified: (1) pre-logging access, (2) felling, (3) skidding and removal of trees and (4) post-clearance site changes.

Pre-logging access means the bulldozing of small access roads adjacent to a grove to be cut, and the movement of equipment into place. Such disturbance, while major, is localized. The soil profile is disrupted, but the greater impact may result from the channelizing of run-off.

The next step is for cutters to move into place and begin the felling operation. Trees are felled uphill so as to minimize their impact, breakage of other trees, and downhill slippage. Nevertheless, felling results in considerable damage to other trees, primarily through the breaking of larger branches. Seedlings and saplings are particularly vulnerable. There is an important relationship between the percentage of trees felled in a stand and the damage which results. UNESCO (Ibid.:458) states that "Felling and logging damage is inevitable. . . . Generally speaking, the severity of damage is correlated more with the number of trees felled than with basal area or volume removed." It is the large crowns of emergent trees which, as they fall, do the greatest amount of damage. Even more damage occurs after trees have been felled. Downed trees must be cut into smaller sizes, removed from the site and loaded onto trucks. The method used within the research area was for sections of four or five meters to be skidded uphill using the winch on the front of the truck, along with a system of pulleys. These pulleys could be easily moved around and suspended from trees to facilitate loading. Bulldozers were not generally used during this phase of the operation.

Of all the activities involved in mechanized logging, skidding has perhaps the greatest impact. Skidding of logs weighing as much as a ton over shallow forest soils causes major disruption of the soil profile. This is made worse by the fact that skidding is generally in an uphill direction. Runoff is thus channeled directly downhill. Skidding also causes physical damage to, or destruction of, seedlings and healthy residuals. Of course the damage to soil and biota is proportional to the degree of felling on a specific forest plot. Within the research area felling operations tended to be very complete, approaching clear-cutting. Damage was thus considerable.

What logging itself does not accomplish in terms of disturbance, site changes following site abandonment do. Most damaging to residuals is microclimatic alteration. Logging generally results in the almost complete removal of the forest canopy. Biotic components are thus directly exposed to insolation and to extremes in climate. It will be recalled that dipterocarp seedlings demand shade and die upon exposure to sunlight. It was claimed by Antonio Rodriguez that for every tree felled in the research area there remained eight healthy residuals. To what extent physical damage has been incorporated into this estimate is not known. Clearly though, these remaining trees were subjected to severe stress. Serevo (1949:373) has observed that:

> young dipterocarps are slender with poorly developed crowns and due to sudden exposure and intense insolation may die eventually. It seems apparent that the residual stand cannot be depended upon to develop to replace the original stand.

Brown and Mathews (1914:457) state that in one case in Bataan, "nearly all seedlings were killed by insolation and most of the small trees became unhealthy and soon died. The few remaining ones seem to stand small chance of reaching maturity." Recall the test plot of 250 square meters where, of 259 dipterocarp seedlings, only 38 survived immediately after logging. Brown and Mathews conclude (Ibid.:537) that:

> The destruction of a portion of these seedlings was probably due to mechanical injuries incident to the logging operation. The great majority, however, have died because of excessive insolation they received during the period just after the main canopy was removed.[35]

Such high rates of mortality mean the destruction of a given forest plot, and set the stage for invasion by secondary species.

The temporal scale of disturbance with logging is generally very short, with the exception of access roads, which must be maintained over an extended period. At any given site felling generally occurs over a period of just a few days or weeks. This clearing activity takes place during the dry season, so immediate erosion is not a problem. Once an area is logged, it is abandoned. Clearance of the ground layer is not continued and invading secondary species begin to become established.

Selection against components of a forest stand occurs at several points in the logging process. Logging activities act directly by removal of certain species, and by damages incident with the processes of road building, cutting, skidding and loading. Following this, post-clearance site changes in microenvironmental elements place severe stress on surviving components. This effect is made even more severe by the strong climatic seasonality characteristic of Zambales. During the dry season insolation is increased tremendously and soils are exposed to serious desiccation. Forest recovery is thus made nearly impossible.

In considering the impact of mechanized logging within the research area, one other factor has played a significant role: in a great number of cases Ayta moved in after particular areas were logged and established swiddens. This was first indicated to me by A. Rodriguez. Subsequent investigation of aerial photographs taken in 1967 show this to be the case; swiddens are clearly visible adjacent to logging

---

35. At higher elevations, approximately 300 meters, they found seedlings of the dipterocarp *Pentacme contorta* surviving after logging. They conclude (Ibid.:457-58) that "The success of *Pentacme* is probably connected with the moist conditions at the high altitude at which the logging is now being done."

roads. Rodriguez estimates that fully 80 percent of those areas logged were subsequently swiddened. This is probably an exaggeration. Clearly however, a large percentage of areas logged were then further cleared for swiddens. Severo (1949) has observed this in other areas of Luzon. Spencer (1966:14-15) notes that:

> As the commercial timberman today is an active cutter of mature forests in may areas of the Orient, he often precedes the shifting cultivator, so that some of the land taken in to the cycle of shifting cultivation is cut over land to start with, containing little but malformed, immature, or less valuable timber.

Such areas would probably make very productive swiddens. As these areas were covered by forest prior to logging, there would be the advantage that soils were yet fertile and their structure intact. Also, much of the work of clearing will have been done--the swiddener only needs to coppice residuals and remove remaining undergrowth. The slash left over from logging activities, once dried and burned, provides nutrients to crops. The primary impact of this pattern of swidden farming following logging would be to destroy any healthy residuals which have not already been killed by insolation.

The parameters of disturbance for this sequence of perturbation would amount to the imposition of swidden disturbance factors over logging disturbance factors. That is, it does not appear that a new set of parameters of disturbance need be invoked to explain or predict the succession which follows. The significance lies in the fact that forest areas which probably would not have been swiddened are brought into this cycle of disturbance.

Mechanized logging has resulted in the complete clearance of numerous stands of timber within the research area. Whether subsequently swiddened or not, these areas have little chance for recovery. As noted, I estimate that 20 to 30 percent of the research area has been logged, most of this since about 1960. Other areas were either already depleted of timber because of swiddening or, like the Kaopan ravine, were too steep and rugged to allow access.

CHAPTER 9

SUCCESSIONAL PROCESSES IN UPLAND ZAMBALES

A. Context, Disturbance, and Initial Floristic Composition

In specific sequences of succession-disturbance-succession the wide range of variables and parameters involved beget a stochastic process of enormous complexity. The goal here has been to view this complexity through the concept of succession, providing both a more parsimonious conceptual model for understanding the history of dynamic ecosystems such as that characteristic of upland Zambales, and an operational model by which to do so. Chapters 3, 4 and 5 were concerned with the physical and biotic context in which successional processes occur in the Cabusilan area. Chapters 6, 7 and 8 identified the three primary types of disturbance which have occurred there--swiddening, burning and logging--and considered the parameters of disturbance--frequency, intensity and scale--for each. The points at which selective pressures are exerted on components of the biotic environment should be clear. It remains to demonstrate how physical, climatic and biotic context and parameters of disturbance together structure successional sequences which commence following disturbance, and thereby structure the course not only of future successional sequences but of subsequent disturbance as well.

An adequate theory of succession should allow us to make certain general statements about the course of, and contingencies within, a given succession. Odum (1971:251), in his definition of succession, notes that:

> It is an orderly process of community development that involves changes in species structure and community

processes with time; it is reasonably directional and, therefore, predictable.

Margalef (1968:30) observes that:

> A theory of succession . . . should pass the test of being explanatory and predictive. But prediction has to be limited to the "macroscopic" level. Prediction in detail is impossible. An ecosystem is a historical construction, so complex that any actual state has a negligible prior possibility.

Colinvaux (1973:88) notes that "The states may be somewhat erratic but within tolerable bounds the successions are predictable." Recognition of certain factors allows us to judge more accurately the range of possibilities under which successional sequences occur. Referring to the biotic element, Drury and Nisbet (1973:332) note that "the theory of succession allows one to predict the characteristics of those species which will replace others, and to recognize unstable or successional species." Certain possibilities can, by this method, be eliminated and the range of possible events narrowed. This is the real utility of an approach which considers the matrix of factors at a given site and the manner in which they are altered by disturbance.

A key to understanding the course of succession on a given site is the recognition that following disturbance, whether mild or severe, a certain portion of the vegetation present before and during disturbance survives. This factor, which has been termed "initial floristic composition"--the living floristic elements present at a given site following disturbance--is of critical importance in determining the course of succession (Egler 1954). Farnworth and Golley (1973:127) note that:

> The significance of remnant populations to succession lies in their potential for rapid establishment at a site early in the succession. If residents are absent, a site may become occupied by an entirely migrant community, effectively suppressing subsequent vegetation changes for a prolonged period.

Attention to this factor is, in effect, a shorthand way of expressing the impact of a given disturbance on the biotic components of a site. Its significance in shaping the course of a future successional sequence cannot be overlooked. Mueller-Dombois and Ellenberg (1974:395) state that:

> The initial floristic composition concept explains certain anomalies that have been observed in secondary succession sequences. . . . There is little doubt that the effect of initial floristic composition (*sensu* Egler) has been underestimated in the interpretation of successional phenomena and that the concept is significant in sharpening the prediction of successional sequences.

Initial floristic composition is both a function of biota existing prior to disturbance and to migratory elements which enter and are established during the period of disturbance. Slatyer (1976:9) notes that Egler (1954)

> suggested that in many cases the "initial floristic composition" following a perturbation may dominate the entire pattern of succession; that is, unless species persisted through the perturbation, or were able to enter the perturbed site shortly afterwards, they would not be represented in the community that developed.

As Slatyer and others have recognized, the concept of initial floristic composition is of critical importance in allowing us to judge among various successional possibilities in particular circumstances.

B. Swiddening

The termination of disturbance following swiddening is not an abrupt event. As harvesting becomes less frequent, invading weeds are not removed with as much diligence. Intermittent harvesting may occur for several years, during which time weeds and other seral species become well established. Thus it is impossible to identify a discrete point at which swiddening ends and succession begins. The cessation of weeding is perhaps the closest approximation.

As discussed, a wide range of vegetational associations may be cleared for swiddens in the Cabusilan area. Of the eight associations defined in Chapter 6 for analysis using SPSS, all but *Imperata* grassland have agricultural potential. Thus, based on vegetational criteria alone, the entire research area except for these grasslands (and high elevation mossy forests) is subject to clearance for swiddens.

The disturbance due to swiddening can be summarized as being of (1) variable frequency, (2) small scale, and (3) high intensity. As noted, these three parameters of disturbance act in different ways. In the context of swiddening, frequency acts selectively by controlling the phenological advancement of component species on a site. Those species which can mature and complete their reproductive cycles before the next clearing are most favored. Repeated frequent disturbance leads to gradual biotic sterilization of sites. At short intervals of clearance woody species are gradually "weeded" out. This accounts for the prevalence of communities of plants such as *amukao* and *Miscanthus* in the research area. Areas in which the intervals are longer develop more stratified successions, beginning with the establishment of softwood trees and ending with mature secondary forest. As species diversity increases through time the result, on subsequent disturbance, is a greater variety of species with potential to survive disturbance and carry on into the next successional sequence.

The intensity of disturbance resulting from swidden clearance is severe. Yet numerous species survive to become components of the initial floristic composition. Primary among these in Zambales are *Miscanthus*, *amukao*, *Chromolaena* and numerous softwood tree species. *Miscanthus*, *amukao* and *Chromolaena* are ubiquitous within the research area and, as noted, possess tenacious root systems. They are effective competitors, particularly the first two, and impede the large-scale invasion of *Imperata*. Depending on soil depth, intensity of burning and other factors, coppiced softwood trees may also survive the swidden cycle. UNESCO (1978:220) notes that "Recovery may . . . occur from root stocks, from coppicing and by other vegetative means. The relative importance of each depends partly on the extent and type of disturbance and the community." The survival of softwood trees may foster a very different kind of succession than might otherwise occur, particularly if these survivors are closely spaced and have the potential to form a light canopy. Light-loving herbaceous plants, including *Miscanthus* and *amukao*, may be of diminished importance in such cases.

As most Ayta swiddens are rather small, the scale of disturbance resulting from swiddening is limited, relative to swidden systems elsewhere in Southeast Asia. This favors a wider variety of propagules, rather than simply grasses and other annuals with efficient seed dispersal mechanisms. When such small former swiddens are adjacent to forest they are more likely to be seeded by species characteristic of primary forest, particularly in later stages of succession. The larger the area of disturbance the more likely it is that that area will be colonized by species possessing long-distance dispersal strategies. Mitigating against the possibility of former swiddens reaching more mature seral stages is the fact that many areas swiddened, particularly on steep slopes, are subsequently subjected to infrequent random burning. This, combined with the mosaic pattern of fallow swidden plots which make early successional disseminules available, maintains large, continuous areas under an early successional regime. The time required for a site to return to "climax" forest, particularly in such a seasonally dry area, is very long. Even "mature" secondary forest containing young members of primary species probably takes on the order to 25 to 40 years. Within the research area the likelihood of an area returning to a climax type of vegetation is thus very low. Once a particular site enters the swidden cycle it would appear that it remains in it. The effect is the transferral to, and maintenance of, a progressively less mature successional biota.

The process of succession is further affected by inherent site factors such as slope. Steeper slopes tend to be more xeric and are thus less hospitable for regenerating vegetation, particularly in such a seasonally dry environment. Also, the position of a particular site on a slope is a factor in the dissemination of propagules. In swiddens cleared on the upper part of slopes or on ridges and outcroppings, particularly in areas where forest has been removed, the probability of K-strategy propagules reaching a plot is relatively low. Plants with more effective dispersal strategies, including most grasses, are much favored. This was noted by Conklin (1959:61), who stated that, "Hilltop and ridge sites remain maximally exposed to light, wind-borne, hairy-glumed grass seed while reseeding from forest stands, the nearest of which are at lower elevation, is minimal." Approximately 50 percent of the swiddens within the research area are cleared from just such sites.

C. Burning

The conditions under which burning occurs within the research area are limited, occurring primarily in *Imperata* grasslands, frequently in *Miscanthus*, and occasionally in mixed stands of *Miscanthus/amukao* and some secondary trees. Fires in other associations are less common. In short, fires generally occur in early successional associations which occur as a result of earlier disturbance.

The disturbance due to burning can be summarized as being of (1) high frequency, (2) low intensity, and (3) large scale. Under a high frequency fire regime only annuals or plants with features bestowing fire-tolerance are able to survive and reproduce. *Imperata*, *Miscanthus*, *amukao* and *Chromolaena* all fit this criterion.

The intensity of disturbance due to burning is low, and selection is thus not as severe as under other disturbance regimes. It is in preventing the establishment of K-selected propagules that the significance of burning lies: this is a function of the frequency of disturbance.

The scale of burning within the research area is variable, but is often considerably larger than disturbance resulting from swiddening and may be of very large scale. As with the parameter of frequency, this functions to prevent colonization by K-strategy disseminules. Only species which employ a strategy of wide-scale dissemination are likely to become established--grasses and other early successional species. The maintenance of grassland is encouraged by the xeric nature of the slopes and ridgetops which they occupy. Not only are grasses generally more tolerant of xeric conditions, but it is such conditions which foster burning. This topographic position favors wind-borne r-strategy disseminules such as those characteristic of various types of grasses.

In recent years the assumption that burning of grasslands actually constitutes a disturbance has been questioned. The most widely held view at present is that the maintenance of grasslands must be seen as a cyclical rather than a lineal phenomenon. Vogl (1977:167) states that:

> A more reasonable approach to grassland succession is to abandon the traditional straight-lined and unidirectional approach and consider it as a cyclical or circular phenomenon. Instead of progressing through a replacement series, occasionally checked and set back by catastrophes, grasslands are maintained as vegetative cycles, the driving forces of these cycles often being fires. Most grassland climates fluctuate from wet to dry and back again on a seasonal, yearly, cyclic or irregular basis, with a growth cycle or response superimposed on these fluctuations. Grasses and associated herbaceous species die back annually or during dry periods, and with the usual slow decomposition rates and resultant fuel accumulations lead to inevitable fires.

Spurr and Barnes (1964:422) remark that:

> from the viewpoint of the fire-dependent community and species, fire is a natural factor whose effects have long been incorporated into species' adaptations and ecosystem dynamics. It is not therefore a disturbance when viewed over the longterm, cyclic history of a fire-dependent ecosystem.

It cannot be denied that the reproductive success of certain species such as *Imperata* are dependent on periodic burning, that these plants create an environment suitable to such burning and that in doing so many other species are eliminated, resulting in the perpetuation of these associations. This is what has been termed a "deflected" succession (Richards 1952:329). It is questionable however whether it is appropriate to regard such a phenomenon as cyclical. To do so is to ignore the ecosystemic properties of such communities. Because a single species may dominate and even guide a successional sequence, does not mean that it is synonymous with that succession. Succession is a community phenomenon, one of interacting individuals, each seeking its own Darwinian fitness, not one of a single dominant species. Numerous studies have shown that under given conditions, notably the cessation of burning, *Imperata* will be shaded out (Brown and Mathews 1914:454). Obviously, something else is going on in this type of succession. Fire is a component of high probability in such a system, but it is still an independent input into the system--biotic components are simply adapted to this high probability, just as *amukao* was originally adapted to a high probability of land slides, tree falls and other natural disturbances. The assumption of a determinant cyclicity of disturbance is teleological, a superfluous attempt to erect an independent set of principles to explain the persistence of grassland ecosystems when basic successional principles will suffice.

D. Logging

The context in which logging in the Cabusilan area may occur is very narrow, limited almost exclusively to culled or unculled primary forest. It appears that there are very few cases in which such areas are allowed to recover after initial disturbance over a sufficient time scale to allow the reestablishment of climax vegetation. In short, once primary forest in this area is felled, there is little chance that it will be allowed to recover. Rather, community composition and structure undergo complete change.

Disturbance by logging is an occurrence of very low frequency. However, following disturbance, the biotic composition is changed such that other types of disturbance are made possible, with the potential for a much higher frequency of subsequent disturbance. An area may be swiddened soon after logging, superimposing another set of disturbance parameters. If grasses, particularly *Imperata*, invade the area, it becomes subject to burning. The potential for an altered frequency regime rests on initial floristic composition, surrounding vegetation and subsequent land use.

Because of the partial or complete removal of the forest canopy which results from logging, the intensity of disturbance is great. Though components of the original forest may remain, the ecosystemic properties of the forest community largely cease to exist. This is particularly the case when swiddening subsequently occurs. The opening of the canopy precipitates growth by fast-growing, light-loving secondary species which may either be already present in the soil or colonize from surrounding areas. Following the swiddening of logged areas, we should expect that species characteristic of primary forest would be largely absent as components of the initial floristic composition, and the chances of mortality for those that remain would be very high. Within the research area, particularly on mesic slopes, *amukao* seems to be the prevalent colonizing species following logging. Numerous areas which have been logged are presently occupied by such associations.

The scale of logging activities is generally large, particularly in comparison to swiddening. Whereas swiddens in the Cabusilan area are dispersed in small patches, logging was more concentrated. It is the size of gaps which is of most significance. Spurr and Barnes (1964:441) remark that "Clear-cutting, or the cutting of large gaps with a diameter of at least twice the height of the stand, will favor the invasion of pioneer species."

As a component of the scale of disturbance, roads served to link logging sites to secondary associations, encouraging colonization by those secondary species. One other aspect of logging served to encourage invasion by r-selected species. As noted, logging roads sought to follow ridges, in areas such as the top of Mt. Balitog. It will be recalled that such areas are more xeric. Also, the dissemination of forest species into such areas is inhibited and wind-borne disseminules encouraged.

E. Factors Affecting Successional Dynamics

Within the Cabusilan area, initial clearance of primary forest begins with swiddening and, after about 1960, logging. Burning is not a component of initial clearance from forest. It is after initial clearance that swiddening and burning maintain early successional communities. Seen in the abstract the process is one of refraction as, through time, successions pass through the "lens" of disturbance. The resulting possibilities for a range of associations to colonize an area and become established are much narrower than they were before disturbance. On an aggregate scale the maturity of the ecosystem is being reduced and subsequently maintained.

Looked at in spatial terms, the result is an ecosystem composed of numerous patches of different age and different floristic composition. The identification of discrete boundaries under such circumstances may be nearly impossible. Those areas not under primary forest can be seen as one continuous ecotone.

Crosscutting all other factors which influence succession within the research area is that of elevation. It is a factor of prime importance in determining various successional possibilities. Most significant, with regard to the agricultural potential of this ecosystem, is the effect of elevation on the relationship between *Imperata* and *Miscanthus*. *Imperata* has a wide elevational tolerance, being able to thrive from sea level to over 2000 meters (though not within the research area). *Miscanthus* on the other hand, occurs in this area only above 500 meters, and can grow at elevations at least as high as *Imperata*. At higher elevations within the research area *Miscanthus* is an effective competitor with *Imperata*, except on ridge areas with dryer edaphic conditions. This occurrence of *Miscanthus* at higher elevations is a primary factor in allowing the Ayta to maintain a productive swidden system. At lower elevations, a considerably smaller percentage of secondary areas are amenable to swiddening.

The relationship between extreme seasonality and the ubiquity of early successional vegetation within the Cabusilan area also requires reiteration. Under a climatic regime such as is characteristic of Zambales, recovery from disturbance is slower due to the seasonally limited availability of water. This was noted by Brown and Mathews (1914:440) early in this century. They state that "The second-growth forest will apparently give place to a dipterocarp one much quicker in a region without a pronounced dry season than in one which has a long season of dry weather." Wharton (1968:159) has noted the "much slower rate of recovery of monsoon or deciduous forest with extended dry seasons." Opler et al. (1977:379) observe that "The intercession of a six-month rainless period in the dry lowlands causes a seasonal pulse to these long-term changes and may slow the overall rate of succession." Not only does this limit growth of individual plants, but also acts selectively against the occurrence of numerous species. Farnworth and Golley (1973:125) note that:

> Succession in tropical areas with highly seasonal rainfall is expected to be quite different from that in regions

with more evenly distributed rainfall, since plants in the early stages will have to survive several months of water shortage.

This is reinforced even more by the prevalence within the research area of numerous steep slopes with shallow soils and xeric conditions. Such conditions particularly favor graminoid species. Of course, seasonal drought also encourages the spread of fire, reinforcing the persistence of early successional vegetation.

The vegetation surrounding the site of a disturbance is of critical importance in determining those species which will appear in a particular successional sequence (Farnworth and Golley 1973:126). This surrounding vegetation is the source of nearly all immigration of disseminules into a plot. Succession in an area surrounded by early secondary species will be very different from succession in areas surrounded by primary forest.

There exists a very important relationship between percentage of surrounding area covered by specific vegetational associations and the probability for those species to propagate within the disturbed area. This relationship can best be expressed as a curve: as the percentage of early secondary vegetation in an area increases arithmetically, its potential for reproducing in a disturbed area increases exponentially. For instance, an area surrounded by 50 percent secondary vegetation and 50 percent primary forest, will not be successfully colonized by forest species. What seeding occurs from primary forest will not result in successful propagules until later stages of succession. This is due to the reproductive strategies and life-history characteristics of component plants in each type of association. It takes very little *Imperata* as a seed source to dominate an area (and subsequently persist), while it requires much more forest, and much greater time, for the latter to dominate. Ricklefs (1973:755) has stated that "Two factors are important in early succession on old fields: (a) the rate of colonization of the habitat by species and (b) changes in the environment caused by colonizing plants themselves." It is in part the element of *rate* of colonization which separates early from late successional species.

This phenomenon has been noted by others. Spencer (1966:40) notes that "As the degree of regional clearing increases, marked patterns of change appear in the sequence of succession of plant-shrub-tree species." Grime (1979:155-56) noting the "trend . . . towards dissected and depauperate floras," observes that:

> One effect of such changes may be to restrict the distribution of late successional plants to 'islands', emigration from which is limited by their characteristically low reproductive output and dispersal efficiencies.

As far back as 1914 Brown and Mathews (1914:441) noted that "If the second growth forest is widely separated from any dipterocarp forest, it is not likely to become dipterocarp in character."

Within the research area an important repository of forest propagules is to be found in depressions and small valleys, usually along streams. These areas, because they enjoy moister edaphic conditions and because they are located in areas generally protected from fire, support large numbers of forest species. Seen from aerial photographs, these areas form the "fern-leaf" pattern mentioned in Chapter 5. Kowal (1966:403), discussing the relation between grassland/*Pinus insularis* and montane forest associations under a fire regime, states that "it is evident that small patches of montane forest in stream depressions play an important role in succession." He continues at length, stating (Ibid.:403-4):

> Practically everywhere in area occupied by pine forest at higher elevations there are numerous stream depressions containing small patches of montane forest. These depressions usually contain less eroded and more organic soil than the adjacent ridges, and thus have a higher soil moisture content. The soil and the depression microclimate result in a favorable habitat for montane forest species. Since the fires lose much of their destructive force when advancing downslope into the depressions, the montane forest patches are relatively immune from fire, even when frequent on the adjacent ridges. These factors are thus responsible for maintaining a widely spread group of isolated montane forest communities potentially capable of initiating the

succession, over a long period of time, of the pine forest into the montane forest, in the absence of fire.

A very similar situation exists within the research area. Remaining forest associations are the primary source of forest propagules. Though swiddens are frequently cleared from such associations (36 percent), they are more protected from subsequent burning, possess more favorable edaphic and microclimatic site conditions, and are more likely to receive K-strategy forest propagules.

F. Assessing the Successional Approach

In this study I have used the concept of succession as a point of focus to provide a framework for defining and explaining the sequence of historical events which has occurred in upland Zambales. It is the particular emphasis upon the parameters of disturbance that initiate successional sequences which to me seems to be of greatest utility in our attempts to understand ecosystems in which humans are primary actors. Such a framework is applicable in a wide range of ecological and ethnographic contexts.

In assessing the utility of a successional approach as developed here, we might ask how this focus on succession and disturbance can increase our understanding of human adaptive processes. It is in defining the reciprocal relationship between ecological circumstances and human responses that the value of such an approach lies. Further, it should allow us to predict the circumstances under which we might expect to observe particular types of responses in the future.

In Zambales the response of Ayta swiddeners to present ecological circumstances can be seen to condition future circumstances. That is, present adaptive strategies can be seen to affect the upland environment in such a way that particular future strategies are either constrained or favored. By their present actions the Ayta are, in a sense, creating future responses. Succession, as developed here, provides a framework for specifying this relationship, both synchronically and diachronically. A careful specification of synchronic circumstances and events allows some statement of both past and future states and processes.

Succession theory provides a framework which is inherently temporal in character. No successional sequence can be divorced from events or sequences antecedent or consequent; the past, present and future of an ecosystem are inextricably linked. Humans, by disturbing these environments, become an integral component of the total ecosystem and of the processes which occur within it. Succession theory provides a moving stage upon which the play of human adaptation is carried out. As such it enables us to define adaptive processes occurring in particular circumstances.

Evidence suggests a long sequence of disturbance in the research area, and from this we can infer human causation. It has in fact been possible to go beyond simple inferences of causation and, within a certain range, to specify the parameters of this disturbance.

Kirch (1980:127) states that "Man frequently simplifies ecosystems (Odum 1969; Margalef 1968), causing them to 'regress' from greater to lesser maturity." The primary ecological process which has occurred through time in upland Zambales has been a regression, and maintenance, of maturity. Forests have been removed at a constant or slowly accelerating rate, while succession leading to forest has not been allowed to resume. Associations characteristic of early succession have persisted through continuous human intervention. Swiddening, orchestrated with burning, has historically been the primary cause of this regression in maturity in upland Zambales. Clearly a long sequence of disturbance has occurred and a temporal change in the parameters of disturbance can be inferred. Logging has been a significant factor within the research area only in relatively recent years.

The scale and frequency of swiddening in upland Zambales has increased very gradually over a long period of time, a particular impetus presumably occurring with the introduction of New World crops. Sweet potato (*Ipomoea batatas*), *Xanthosoma*, corn (*Zea mays*), cassava (*Manihot esculenta*) and lima beans (*Phaseolus lunatus*) have been of greatest significance in this respect. Pigeon pea (*Cajanus cajan*), of recent introduction from Africa, has also been important. The factor which sets these crops apart is their very high return for minimal amounts of labor input, their ability to produce well in soils of relatively low fertility, and their drought resistance. Not only have high yields reduced one limitation on population growth but, as subsequent expansion of swiddening and disturbance resulted in changed edaphic and biotic conditions, these crops have been able to continue producing at high levels. These are crops which have what Harlan (1972) has called a wide "ecological amplitude." Such crops, he notes (Ibid.:239) "can extend the range of agricultural settlement in several directions." Sweet potatoes alone, he observes (Ibid.:241), can support very dense populations; up to 200-250 individuals per sq. kilometer in highland New Guinea, for instance. Crops such as taro, yams and bananas are much more selective in their requirements, and do not produce as well under less optimal conditions.

While the scale of disturbance from swiddening has changed only gradually through time, from smaller to larger swiddens, the frequency of disturbance, on a regional scale, has increased dramatically. Initially this would have followed on the tails of swidden clearance but, with time, would have generated its own set of disturbance parameters, particularly of scale and frequency: both of these can be seen as increasing greatly over time. During the early period of swidden clearance, vegetational associations subject to burning would presumably have been of much more patchy distribution. This lack of spatial continuity would seriously impede the spread of fires. As clearance of forest continued through time, continuity would have become much greater. This would also allow a continually higher frequency of burning, simply because of the increased probability of fire occurring over increasingly larger areas and because of the increased quantity of r-strategy propagules available.

The net result of this historical sequence has been the creation of an environment allowing continued agricultural activity, i.e. an agroecosystem. Hardesty (1977:95), paraphrasing Odum (1971), states that:

> it is to nature's advantage for an ecological system to increase gross production while it is to mans' advantage to increase net production. One way of increasing net production is to manipulate the process of ecological succession.

Margalef (1968:47) has noted that "Human exploitation in a sustained form is possible only if carried on in an ecosystem of low maturity, with a relatively simple structure and high ratio of primary production to total biomass."

It remains to consider future contingencies in the maintenance of the upland agroecosystem created by the Ayta. The present agricultural system in the Cabusilan area is clearly sustainable and has the potential for even further expansion. It has been demonstrated that early successional vegetational associations such as *Miscanthus* and *amukao* make productive swiddens and, due in large part to inherent site factors--particularly elevation--are effective competitors with *Imperata* and other less desirable species. Such communities, being effective competitors, persist even under repeated disturbance. An environment is being created which is capable, by the nature of the secondary vegetation and the factors which favor it, of supporting continued swidden activity. Brookfield (1968:417) states that "We would assess agriculture in any given locality from the point of view of its efficiency in producing food without damage to the resources available to future generations on the same land" (Brookfield and Brown 1963:166-67). Forest clearance can, and will, continue for some time, presumably for as long as any such forest remains. Ultimately, pushed to the limits, we might hypothetically envisage an ecological scenario in which the flora of the Cabusilan area consists of little other than *Imperata*, *Miscanthus*, *amukao* and scattered stands of fast growing trees. Yet even under this extreme scenario (and it is not anticipated that this will in fact occur) productive agriculture can persist. Fallow periods can probably be shortened considerably, as they already have been in some particularly desirable or accessible areas. At such a hypothetical state the gross productivity of human agricultural production will be at its highest point, though forest will have disappeared. It is not necessary for fallow plots to return to forest to maintain productivity. The Ayta have created a patchwork of stable truncated successions which are being maintained, and which seem capable of remaining productive for many years to come.

In the evolution of the present system a crop complex has been selected (and, in the Darwinian sense, selected for) which will continue to be adaptive in the steady provision of carbohydrates and vegetable proteins. Crops such as corn and cassava, which are particularly tolerant of xeric conditions, may become increasingly important, particularly at lower elevations. Here, ultimately, extreme seasonality is the greatest limiting factor. Sweet potato and

Xanthosoma, which do particularly well at higher elevations, will probably remain unchallenged as the primary sources of carbohydrates. Pigeon pea and lima bean are likewise certain to be maintained.

CHAPTER 10

CONCLUSION

The history of an ecosystem is more than a series of events which can only be described in the particular. The preceding discussion has been an attempt to assess the historical sequence of changing ecosystemic relations in upland Zambales from the perspective of succession theory. By focusing both on ecological context and on disturbance, it has been possible to make some sense of the patterning of successions observable within the research area and to incorporate particular historical events into a broader theoretical context. It is the particular emphasis upon the parameters of disturbance which initiate successional sequences which allows us to more parsimoniously explain the genesis of ecosystems in which humans are primary actors.

Within the research area a number of contextual factors have played prominent roles in the structuring of past and present successional sequences. Primary among these has been the extreme seasonality of rainfall characteristic of the area. The rugged physiography, high percentage of steep slopes over total land area, and differences in elevation have also been significant in ways previously specified. Finally, the floral portfolio available for both clearance and colonization, both endemic and introduced, have been of critical importance. *Miscanthus*, *amukao*, *Chromolaena odorata* and many species of softwood secondary trees, notably *Ficus spp.*, have played major roles in vegetational dynamics. These contextual factors, along with the parameters of disturbance--frequency, intensity, and scale--have brought about the present biotic environment. In effect, disturbance has, over time, led to changes in the context of which further disturbance occurs. The three primary types of anthropogenic disturbance which have occurred in this area are swiddening, burning and logging.

It it clear that much of the deforestation which has occurred in upland Zambales is the result of clearance by Ayta for swidden fields over a period of several centuries. Most available descriptions of Ayta swiddening tend to characterize it as desultory and unsophisticated. This is a highly inaccurate characterization. Ayta are, as noted, a population firmly and fully committed to swidden agriculture. Hunting and foraging activities are in fact less important here than among many other Philippine swidden societies which have been described (Schlegel 1979; Wallace 1970). Certainly this has not always been the case. I have suggested that three phases in the historical sequence of Ayta subsistence adaptation can be recognized: (1) hunting-gathering, (2) pre-Hispanic swiddening, and (3) post-Hispanic swiddening, characterized by the introduction of New World crops. The present Ayta adaptation represents an adjustment to marked environmental degradation, as can be inferred from historical, ecological and other evidence. The process has, of course, been reciprocal; clearly much of this environmental change which has occurred is the result of Ayta activity. Swiddening, orchestrated with widespread burning, is largely responsible for the creation of the mosaic environment observed today. The greater part of the research area, and much of the rest of southern Zambales, is today covered with early secondary vegetation. The pay-off for this has been increased production and a greater degree of self-sufficiency. By creating and maintaining a less mature environment the Ayta have established an economically more productive habitat. It appears to be an agroecosystem which is sustainable with continued high productivity.

This brings into sharp relief the situationality of particular swidden systems with regard to the question of impact and sustainability. Researchers have tended to think and write polemically on this question. Anthropologists tend to stress the adaptive aspects of such systems, while foresters, administrators, and others emphasize their destructiveness. Rarely has the issue been considered ecosystemically and, until recently, never from the perspective of succession. Every swidden system is different: considerations of impact and sustainability can be addressed only with reference to a set of factors specific to each system. These are factors of (a) the physiographic, climatic and biotic context, and (b) the parameters of disturbance. Without making these explicit, it is impossible to accurately judge the sustainability of a particular swidden system. Superficially similar systems may differ widely in their impact by as minor a factor as the presence or absence of a single successional plant species.

It is clear that in upland Zambales swidden farming by Ayta has resulted in considerable deforestation. I would not presume to act as an apologist for Ayta in this matter, nor is it a matter of laying blame. The impact of Ayta swiddening on the primary forest communities is probably due more to the extreme seasonality characteristic of western Luzon than any other factor. More important, the preceding should not be read as a statement of the destructiveness of swidden agriculture. In fact the Ayta system seems to stand out among other swidden systems in this sense.

The requirements imposed by a successional perspective have made it possible to disentangle those factors which have had relevance in the genesis and formation of the present upland ecosystem in Zambales, an ecosystem in which the Ayta have been primary actors. They continue to play a part in the maintenance and creation of ecosystemic relations within this environment. Present circumstances have their foundation in past events. And by their present actions the Ayta are creating future responses.

APPENDIX A

SPSS FORMAT AND SWIDDEN DATA

Definition of Format

File name: SWIDDEN

Variable labels:

AGE- Designates age of swidden. 0= newly cleared, i.e., 1980 dry season; 1= cleared 1979 dry season; 2= cleared 1978 dry season.

EL- Elevation of swidden above sea level measured from middle of swidden. Stated in meters.

SZ- Size of swidden in hectares.

LOC- Location of swidden in one of three survey areas. 1= Maagə'agə', 2= Kalantas, 3= Bayabas. See map for locations.

DIR- Aspect of swidden. Expressed in degrees on compass. 90°= east, 180°= south, 270°= west, 360°= north.

MIN- Minimum inclination of swidden.

MAX- Maximum inclination of swidden.

CROP 1- Sweet potato, *Ipomoea batatas.*

CROP 2- *Xanthosoma sagittafolium.*

CROP 3- Corn, *Zea mays.*

CROP 4- Taro, *Colocasia esculenta.*

CROP 5- Rice, *Oryza sativa.*

CROP 6- Banana, *Musa sapientum, M. paradisiaca.*

CROP 7- Cassava, *Manihot esculenta.*

CROP 8- Yam, *Dioscorea alata.*

CROP 9- Pigeon pea, *Cajanus cajan.*

CROP 10- Lima bean, *Phaseolus lunatus.*

CROP 11- *Bagoyboy, Thysanolaena maxima.*

PREVEG 1- Grassland; Includes members of genera *Imperata, Saccharum* and *Rottboelia.*

PREVEG 2- *Miscanthus.*

PREVEG 3- *Amukao.*

PREVEG 4- Young secondary forest; *Ficus* predominant.

PREVEG 5- *Miscanthus/Amukao.*

PREVEG 6- *Miscanthus/Amukao*/young secondary forest association.

PREVEG 7- Mature secondary forest.

PREVEG 8- Primary forest.

PREVEG 9- Riparian vegetation; has elements of PREVEG 4, 7 and 8, but distinct floristic composition as well.

Value labels: For CROP 1- CROP 11 & PREVEG 1- PREVEG 9.

1= Present, but few represented.

2= Present, many represented.

3= Co-dominant. Present in roughly equal numbers to one or more other cultigens.

4= Dominant, but other cultigens plentiful.

5= Heavily dominant. Nearly exclusive, but other cultigens present in small numbers.

6= Exclusive. No other cultigens present.

Methods

FREQUENCIES: Subprogram FREQUENCIES was used in order to generate one-way frequency distribution tables, histograms and descriptive statistics. It was employed for the following variables:

AGE
EL
DIR
LOC
MIN
MAX
SZ
CROP 1-11
PREVEG 1-9

CROSSTABS: Subprogram CROSSTABS was employed in order to investigate relationships between several sets of variables. Contingency tables were generated along with summary statistics and measures of association. The following variables were tested with the CROSSTABS procedure:

CROP 1-11/PREVEG 1-9
CROP 1-11/CROP 1-11
CROP 1-11/AGE
CROP 1-11/EL
CROP 1-11/DIR
CROP 1-11/SZ
CROP 1-11/LOC
CROP 1-1/MIN
CROP 1-11/MAX
PREVEG 1-9/PREVEG 1-9
PREVEG 1-9/EL
PREVEG 1-9/DIR
PREVEG 1-9/SZ
PREVEG 1-9/LOC
PREVEG 1-9/MIN
PREVEG 1-9/MAX
LOC/SZ
EL/SZ
MIN/SZ
MAX/SZ

NONPAR CORR: Subprogram NONPAR CORR was used for the purpose of generating Spearman correlation coefficient matrices for the analysis of CROPS and PREVEG, which are scaled on an ordinal level. Three tables were generated in this manner:

CROP 1-11/CROP 1-11
CROP 1-11/PREVEG 1-9
PREVEG 1-9/PREVEG 1-9

In addition to the above, diversity indices were calculated for CROP 1-11 and PREVEG 1-9 for all 39 swidden cases, and plotted on several scattergrams. This procedure was done "by hand." The following were plotted:

Crop diversity indices/Preveg diversity indices
Crop diversity indices/Elevation
Crop diversity indices/Size

Case #	Age	Elevation (Meters)	Size	Location	Aspect	Min. Slope	Max. Slope	Crops											Preveg								
								1	2	3	4	5	6	7	8	9	10	11	1	2	3	4	5	6	7	8	9.
19	1	950	.15	3	90	30	40	6	0	0	0	0	0	0	0	0	0	0	0	0	6	0	0	0	0	0	0
20	0	600	.50	1	360	40	40	0	2	3	0	3	0	0	0	1	1	0	0	0	0	0	0	0	2	4	0
21	0	430	.80	1	225	00	50	0	1	3	0	3	0	0	1	1	1	1	1	0	0	0	5	0	0	0	0
22	1	470	.30	1	180	15	25	0	0	3	0	3	0	3	0	2	0	3	0	0	6	0	0	0	0	0	0
23	0	420	1.50	1	315	00	50	0	1	3	1	4	1	0	1	1	0	1	1	3	2	0	0	3	0	0	2
24	0	400	.50	1	240	20	20	0	0	3	0	3	0	0	0	0	1	0	0	0	5	1	0	0	0	0	0
25	0	740	.25	2	25	25	25	0	2	4	0	0	0	0	0	0	0	0	0	0	0	0	0	6	0	0	0
26	0	740	1.20	2	45	00	45	0	0	3	0	3	0	0	0	0	0	0	0	3	3	2	0	0	0	0	0
27	0	740	.25	2	270	00	45	3	2	3	1	0	2	0	1	0	0	0	0	0	0	0	0	4	2	0	0
28	0	920	.15	3	150	00	10	6	0	0	0	0	0	0	0	0	0	0	0	0	0	0	0	6	0	0	0
29	0	610	.75	1	315	5	35	1	2	3	1	3	0	1	1	1	2	0	0	0	0	0	3	4	0	0	0
30	0	580	.15	1	200	40	40	0	3	3	2	0	0	0	0	1	1	2	0	0	0	0	0	0	6	0	0
31	0	580	.25	1	135	30	60	0	3	3	0	0	0	0	1	1	2	0	0	0	0	0	3	0	3	0	3
32	1	600	.20	1	335	10	15	5	1	0	0	0	0	0	0	0	0	0	0	0	0	0	4	0	2	0	0
33	1	600	25	1	225	30	35	0	0	3	0	3	0	1	0	1	2	2	0	5	0	1	0	0	0	0	0
34	0	660	.25	2	315	30	50	3	3	3	0	0	2	0	1	0	0	0	0	0	0	0	0	0	0	6	0
35	2	740	.20	2	000	00	00	0	3	0	0	0	3	0	1	0	1	0	0	0	0	0	0	0	6	0	0
36	1	760	.20	2	000	00	10	2	0	0	0	0	1	1	0	2	0	1	0	0	0	0	0	6	0	0	0
37	0	760	.25	2	20	00	15	3	3	3	0	3	0	0	1	0	1	0	0	0	0	0	0	6	0	0	0
38	1	590	.25	1	180	10	30	3	0	0	0	0	0	3	0	0	0	2	0	6	0	0	0	0	0	0	0
39	2	740	.20	2	40	20	20	4	2	0	0	0	0	0	0	2	1	0	0	0	0	0	0	6	0	0	0

Swidden Data

Case #	Age	Elevation (Meters)	Size	Location	Aspect	Min. Slope	Max. Slope	Crops 1	2	3	4	5	6	7	8	9	10	11	Preveg 1	2	3	4	5	6	7	8	9
1	0	950	.25	3	135	10	55	5	2	0	2	0	0	0	0	0	0	0	0	0	0	0	0	0	0	6	0
2	0	920	.15	3	145	20	40	3	3	3	3	0	0	0	0	0	0	0	0	0	0	0	0	0	0	6	0
3	0	920	.20	3	135	10	30	3	2	3	1	0	0	0	1	1	1	0	0	0	0	0	0	0	0	6	0
4	1	920	.20	3	360	00	50	5	2	4	2	0	0	0	0	1	1	0	0	4	0	0	0	2	0	0	0
5	0	920	.15	3	360	00	70	4	3	3	1	0	1	0	0	1	1	0	0	3	0	0	0	0	0	0	3
6	1	1000	.60	3	20	35	35	5	2	0	1	0	0	0	0	0	1	0	0	2	0	0	0	0	0	4	0
7	0	1020	.20	3	315	45	50	5	1	0	0	0	0	0	0	0	0	0	0	0	0	0	0	0	0	6	0
8	1	1020	.20	3	25	20	55	6	0	0	0	0	0	0	0	0	0	0	0	0	0	0	0	0	0	6	0
9	0	930	.25	3	150	45	55	3	2	3	1	0	0	0	0	0	0	0	0	0	0	0	3	0	0	0	3
10	1	950	.40	3	150	20	40	5	1	0	0	0	0	1	0	2	1	2	0	0	0	0	3	0	3	0	0
11	1	935	.25	3	150	15	30	5	1	0	0	0	0	0	0	0	0	2	0	0	0	0	0	5	0	1	0
12	0	970	.25	3	90	40	45	3	3	3	2	0	0	0	0	1	1	0	0	0	0	0	0	6	0	0	0
13	1	970	.20	3	145	20	40	5	2	0	2	0	0	1	0	1	1	1	0	0	0	0	0	6	0	0	0
14	1	1000	.25	3	180	30	30	5	1	0	1	0	0	0	1	1	0	0	0	0	0	0	0	0	0	6	0
15	0	1020	.25	3	45	45	50	4	3	0	1	0	0	0	0	0	0	0	0	0	0	0	0	0	6	0	0
16	1	1020	.20	3	360	45	45	4	2	0	2	0	0	0	0	3	1	1	0	0	0	0	0	0	6	0	0
17	1	910	.25	3	270	15	35	5	1	0	1	0	0	0	0	0	0	1	0	0	0	0	0	0	6	0	0
18	0	950	.15	3	270	30	40	3	3	3	0	0	0	0	0	0	0	0	0	0	0	0	0	6	0	0	0

APPENDIX B

SPEARMAN CORRELATION COEFFICIENTS

	Grassland	Miscanthus	Amukao	Young sec.	Miscanthus/ Amukao	Miscanthus/ Amukao/ Young sec.	Mature sec. forest	Primary forest	Riparian vegetation
Sweet potato	-0.2971 .033	-0.1105 .252	-0.2481 .064	-0.3687 .010	-0.0987 .275	0.0368 .412	-0.0769 .321	0.3016 .031	-0.1905 .123
Xanthosoma	-0.1390 .199	-0.2067 .103	-0.4738 .001	-0.3979 .006	-0.0007 .498	0.1204 .233	0.2864 .039	0.0510 .379	0.2122 .097
Corn	0.1864 .128	0.1999 .111	0.1538 .175	0.2312 .078	0.0747 .326	0.1229 .228	-0.2105 .099	-0.1724 .147	0.2707 .048
Colocasia	-0.0115 .472	0.0635 .350	-0.2137 .096	-0.2435 .068	-0.1452 .189	-0.0083 .480	0.1059 .261	0.1066 .259	0.1139 .245
Rice	0.4724 .001	0.2399 .071	0.5202 .001	0.5074 .001	0.1071 .258	-0.0259 .438	-0.2069 .103	-0.2041 .106	0.0378 .410
Banana	0.1975 .114	0.1237 .227	0.0125 .470	-0.1225 .229	-0.1805 .136	0.0868 .300	0.0956 .281	-0.0653 .347	0.2758 .045
Cassava	-0.1083 .256	0.1881 .126	0.0630 .352	0.0963 .280	0.1330 .210	0.0810 .312	-0.1347 .207	-0.2710 .048	-0.1573 .169
Ubi	0.3959 .006	-0.1287 .217	-0.0674 .342	-0.1694 .151	0.2413 .069	0.0219 .447	0.0443 .394	0.0888 .295	0.1785 .139
Pigeon pea	0.1840 .131	0.0386 .408	-0.0173 .458	-0.0993 .274	0.1516 .178	0.0910 .291	0.0863 .301	-0.2245 .085	0.1217 .230
Lima bean	0.0000 .500	0.1197 .234	-0.2068 .103	0.1622 .162	0.2456 .066	-0.0062 .485	0.1649 .158	-0.2339 .076	0.0859 .302
Bagoyboy	0.2664 .051	0.1689 .152	0.0979 .277	0.0361 .414	0.0270 .435	-0.0509 .379	0.1451 .189	-0.2735 .046	-0.0808 .312

	Sweet potato	Xanthosoma	Corn	Colocasia	Rice	Banana	Cassava	Ubi	Pigeon pea	Lima bean	Bagoyboy
Sweet potato		-0.1462 .187	-0.6482 .001	0.1732 .146	-0.6005 .001	-0.2277 .082	-0.1442 .191	-0.3243 .022	-0.2248 .084	-0.2975 .033	-0.1603 .165
Xanthosoma			0.3124 .026	0.4362 .003	-0.2882 .038	0.1876 .126	-0.3784 .009	0.2646 .052	-0.0108 .474	0.3124 .026	-0.3746 .009
Corn				0.0990 .274	0.4375 .003	0.0713 .333	-0.1169 .239	0.2530 .060	0.1079 .257	0.2570 .057	-0.1598 .166
Colocasia					-0.2507 .062	-0.0326 .422	-0.1424 .194	-0.0292 .430	0.1715 .148	0.1205 .232	-0.0392 .406
Rice						-0.0456 .391	0.2098 .100	0.2529 .060	0.1836 .132	0.2537 .060	0.1837 .131
Banana							-0.0370 .412	0.4240 .004	0.0065 .484	-0.1211 .231	-0.0428 .398
Cassava								-0.1290 .217	0.4118 .005	0.1492 .182	0.6106 .001
Ubi									0.0697 .337	0.2117 .098	-0.1730 .146
Pigeon pea										0.5306 .001	-0.4333 .003
Lima bean											0.0479 .386
Bagoyboy											

REFERENCES

Agpaoa, A., et al.
1975 *Manual of Reforestation and Erosion Control for the Philippines*. Stuttgart: German Agency for Technical Cooperation, Ltd.

Alcala, A. C.
1976 *Philippine Land Vertebrates*. Quezon City: New Day Publishers.

Alicante, M. M., et al.
1951 *Soil Survey of Zambales Province, Philippines*. Department of Agriculture and Natural Resources, Soil Report 13. Manila: Bureau of Printing.

Bartlett, H. H.
1956 "Fire, Primitive Agriculture and Grazing in the Tropics." In *Man's Role in Changing the Face of the Earth*, W. L. Thomas, Jr. (ed.). Chicago: University of Chicago Press.

Bennett, J. W.
1973 "Ecosystemic Effects of Extensive Agriculture." *Annual Review of Anthropology*, V. 2. Palo Alto: Annual Reviews, Inc.

1975 "Ecosystem Analogies in Cultural Ecology." In *Population, Ecology and Social Evolution*, Steven Polgar (ed.). The Hague: Mouton Publishers.

Blair, E. H., and J. A. Robertson (eds.)
1909 *The Philippine Islands, 14-93-1898*, 55 Vols. Cleveland: A. H. Clark Co.

de Boer, H. H.
1950 "On the Relation between Rainfall and Altitude in Java, Indonesia." *Chronica Naturae* 106(8-9):424-427.

Bragge, T. B., and L. C. Hulbert
1976 "Woody Plant Invasion of Unburned Kansas Bluestem Prairie." *Journal of Range Management* 29:19-24.

Brookfield, H. C.
1968 "New Directions in the Study of Agricultural Systems in Tropical Areas." In *Evolution and Environment*, E. T. Drake (ed.). New Haven: Yale University Press.

Brookfield, H. C., and P. Brown
1963 *Struggle for Land: Agriculture and Group Territories among the Chimbu of the New Guinea Highlands*. Melbourne: Oxford University Press.

Brown, W. H.
1919 *Vegetation of Philippine Mountains*. Publication No. 13. Department of Agriculture and Natural Resources. Manila: Bureau of Printing.

Brown, W. H., and D. M. Mathews
1914 "Philippine Dipterocarp Forests." *Philippine Journal of Science* 9(5):413-561.

Budowski, G.
1966 "Fire in Tropical American Lowland Areas." In *Proceedings Fifth Annual Tall Timbers Fire Ecology Conference*. Tallahassee, Fla.

Bureau of Mines, Philippines.
1963 *Geological Map of the Philippines*. Manila: Bureau of Mines.

Butzer, K. W.
1974 "Accelerated Soil Erosion: A Problem of Man-Land Relationships." In *Perspectives on Environment*, I. R. Manners and M. W. Mikesell (eds.). Publication No. 13. Washington, D.C.: Association of American Geographers.

de Candolle, A.
1855 *Geographie Botanique Raisonnee; ou Exposition des Faits Principaux et des Lois Concernant la Distribution Geographique des Plantes l'epoque Actuelle*. 2 Vols. Paris: V. Masson.

Chapin, F. S., and K. Van Cleve
n.d. *Plant Nutrient Absorbtion and Retention under Differing Fire Regimes*. Unpublished paper. In files of Environment and Policy Institute, East-West Center, Honolulu.

Chief of Naval Operations, Aerology Section
1944 *Climate and Weather of the North and Central Philippine Islands*. June 1944.

Clements, F. E.
1916 *Plant Succession: An Analysis of the Development of Vegetation*. Washington: Carnegie Institute Publications, No. 242.

Colinvaux, P.
1973 *Introduction to Ecology*. New York: John Wiley and Sons, Inc.

1978 *Why Big Fierce Animals are Rare: An Ecologist's Perspective*. Princeton, New Jersey: Princeton University Press.

Conklin, H. C.
1954 "An Ethnoecological Approach to Shifting Agriculture." *Transactions of the New York Academy of Sciences*, 2nd Series, 17:133-142.

1957 *Hanunoo Agriculture*. Rome: Food and Agriculture Organization.

1959 "Shifting Cultivation and Succession to Grassland." In *Proceedings, Ninth Pacific Science Congress*. Bangkok.

Coronas, J.
1920 *Climate and Weather of the Philippines, 1903-1918*. Manila: Bureau of Printing.

Cortes, R. M.
1974 *Pangasinan, 1572-1800*. Quezon City: University of the Philippines Press.

Daubenmire, R.
1968 "Ecology of Fire in Grasslands." In *Advances in Ecological Research*, J. B. Cragg (ed.). Vol. 5. London: Academic Press.

Department of Natural Resources (Philippines)
n.d. *Philippine Mammals: Nature's Treasures*. Quezon City: Department of Natural Resources.

Dickerson, R. E.
1928 *Distribution of Life in the Philippines*. Manila: Bureau of Printing.

Diong, C. H.
1973 "Studies of the Malayan Wild Pig in Perak and Johore." *Malayan Nature Journal* 26:120-51.

Drury, W. H., and I. C. T. Nisbet
1973 "Succession." *Journal of the Arnold Arboretum* 54(3):331.

Egler, F. E.
1954 Vegetation Science Concepts. 1. Initial Floristic Composition -- A Factor in Old Field Vegetation Development." *Vegetation* 4:412-17.

Estioko-Griffin, A. A., and P. B. Griffin
1975 "The Ebuked Agta of Northeastern Luzon." *Philippines Quarterly of Culture and Society* 3(4):237-44.

1981 "Woman the Hunter: The Agta." In *Woman the Gatherer*, Francis Dahlberg (ed.). New Haven: Yale University Press.

Eussen, J. H. H., et al.
1976 "Competition Between Alang-Alang (*Im. cyl.* (L.) Beauv.) and Some Crop Plants." *Biotrop Bulletin*, No. 10. Bogor, Indonesia: BIOTROP, SEAMEO. Regional Center for Tropical Biology.

Evangelista, A. E., and A. T. Madale
1961 "Digging Up History: Recent Archaeological Surveys in Marinduque and Zambales Bare Data on Philippine Past." *Sunday Times Magazine* (Manila), April 16, 1961.

Farnworth, E. G., and F. B. Golley (eds.)
1973 *Fragile Ecosystems*. New York: Springer-Verlag.

Flenley, J.
1979 *The Equatorial Rain Forest: A Geological History*. London: Butterworths.

Flores, J. F., and V. F. Balagot
1969 "Climate of the Philippines." In *Climates of Northern and Eastern Asia*, H. Arakawa (ed.). World Survey of Climatology, V. 8. Amsterdam: Elsevier Publishing Co.

Fox, R. B.
1953 "The Pinatubo Negritos: Their Useful Plants and Material Culture." *Philippine Journal of Science* 81(3-4).

Frake, C. O.
1960 "The Eastern Subanun of Mindanao." In *Social Structure in Southeast Asia*, G. P. Murdock (ed.). Viking Fund Publications in Anthropology, No. 29. Chicago: Quadrangle Books.

1962 "Cultural Ecology and Ethnography." *American Anthropologist* 64:53-59.

Freeman, D.
1970 *Report on the Iban*. London School of Economics and Political Science, Monographs on Social Anthropology No. 41. London: Athlone Press.

Gams, H.
1918 "Prinzipienfragen der Vegetationforschung. Ein Beitrag zur Begriffskirung and Methodik der Biocoenologie." *Vierteljahrsschr. Naturforsch. Ges. Zurich* 63:292-493.

Garvan, J. M.
1963 *Negritos of the Philippines*. Horn-Wien: Verlag Ferdinand Berger.

Geertz, C.
1963 *Agricultural Involution: The Processes of Ecological Change in Indonesia*. Berkeley: University of California Press.

Gemelli Careri, G. F.
1963 *A Voyage to the Philippines*. Manila: Filipiniana Book Guild.

Gervasio, F. C.
1966 "The Age and Nature of Orogenesis of the Philippines." *The Philippine Geologist* 20(4):121-141.

Goody, J.
1958 *The Developmental Cycle in Domestic Groups*. Cambridge: Cambridge University Press.

Gourou, P.
1956 "The Quality of Land Use of Tropical Cultivators." In *Man's Role in Changing the Face of the Earth*, W. H. Thomas, Jr. (ed.). Chicago: University of Chicago Press.

1958 *The Tropical World*. London: Longman.

Grime, J. P.
1979 *Plant Strategies and Vegetation Processes*. Chichester: John Wiley and Sons.

Grubb, P. J.
1977 "Control of Forest Growth and Distribution of Wet Tropical Mountains: With Special Referral to Mineral Nutrition." *Annual Review of Ecology and Systematics* 8:83-107. Palo Alto: Annual Reviews, Inc.

Guido, J. P.
1916 "The Negritos near Botolan, Zambales." Negrito-Aeta Paper No. 11. In *Ethnography of the Negrito-Aeta Peoples, A Collection of Original Sources*, H. O. Beyer (ed.). Manila, 1918.

Hames, R. B.
1979 "A Comparison of the Efficiencies of the Shotgun and the Bow in Neotropical Forest Hunting." *Human Ecology* 7(3):219-52.

Hardesty, D. L.
1977 *Ecological Anthropology*. New York: John Wiley and Sons.

Harlan, J.
1972 "Crops that Extend the Range of Agricultural Settlement." In *Man, Settlement and Urbanism*, P. J. Ucko, R. Tringham and G. W. Dimbleby (eds.). London: Gerald Duckworth and Company, Ltd.

Harris, D. R.
1972 "Swidden Systems and Settlement." In *Man, Settlement and Urbanism*, P. J. Ucko, R. Tringham and G. W. Dimbleby (eds.). London: Gerald Duckworth and Company, Ltd.

Hatch, E.
1973 "The growth of economic, subsistence, and ecological studies in American anthropology." *Journal of Anthropological Research* 29:221-43.

Holm, L. G. et al.
1977 *The World's Worst Weeds: Distribution and Biology*. Honolulu: University Press of Hawaii.

Horn, H. S.
1974 "The Ecology of Secondary Succession." *Annual Review of Ecology and Systematics* 5:25-37.

Horton, R. E.
1945 "Erosional Development of Streams and their Drainage Basins." *Bulletin of the Geological Society of America* 56:275-370.

Hubbell, S. P.
1979 "Tree Dispersion, Abundance, and Diversity in a Tropical Dry Forest." *Science* 203(4387):1299-1309.

Hudson, N. W.
1971 *Soil Conservation*. London: Batsford.

Hudson, N. W., and D. C. Jackson
1959 "Results achieved in the Measurement of Erosion and Runoff in Southern Rhodesia." In *Proceedings, Third Inter-African Soils Conference*, Dalaba CCTA:575-83.

Huke, R. E.
1963 *Shadows on the Land: An Economic Georgraphy of the Philippines*. Manila: Bookmark.

Humboldt, A. von
1805 *Essay sur la Geographie des Plantes*. Paris: Levrault, Schoell et Cie.

IRRI (International Rice Research Institute)
1974 *An Agro-Climatic Classification for Evaluating Cropping Systems Potentials in S.E. Asian Rice Growing Regions*. Los Banos, Laguna, Philippines: IRRI.

Jackson, I. J.
1977 *Climate, Water and Agriculture in the Tropics*. London: Longman.

Janzen, D. H.
1970 "Herbivores and the Number of Tree Species in Tropical Forests." *American Naturalist* 104:501-28.

1973 "Tropical Agroecosystems." *Science* 182:1212-19.

1979 "How to be a Fig." *Annual Review of Ecology and Systematics* 10:13-51. Palo Alto: Annual Reviews, Inc.

Kellman, M. C.
1969 "Some Environmental Components of Shifting Cultivation in Upland Mindanao." *Journal of Tropical Geography* 28:40-56.

Kintanar, R. L.
1973 *Climatology and Wind Related Problems in the Philippines*. Quezon City: PAGASA (Weather Bureau).

Kirch, P. V.
1980 "The Archaeological Study of Adaptation: Theoretical and Methodological Issues." In *Advances in Archaeological Method and Theory*, Vol. 3. New York: Academic Press.

Komarek, E. V., Sr.
1967 "Fire--and the Ecology of Man." In *Proceedings, Sixth Annual Tall Timbers Fire Ecology Conference*. Tallahassee, Fla.

Kostrowicki, J.
1965 "Land Utilization; Case Studies; Origins, Aims, Methods, Techniques." In *Land Utilization in East-Central Europe: Case Studies*, V. Kostrowicki (ed.). Warsaw: Geographica Polonica, 5. Wydawnietua Geologiezue.

Kowal, N. E.
1966 "Shifting Cultivation, Fire, and Pine Forest in the Cordillera Central, Luzon, Philippines." *Ecological Monographs* 36:389-419.

Kucera, C. L.
n.d. *Grasslands and Fire*. Unpublished paper. In files of Environment and Policy Institute, East-West Center.

Larkin, J. A.
1972 *The Pampangans*. Berkeley: University of California Press.

Leigh, E. G., Jr.
1975 "Structure and Climate in Tropical Rain Forest." *Annual Review of Ecology and Systematics* 6:67-86. Palo Alto: Annual Reviews, Inc.

Margalef, R.
1968 *Perspectives in Ecological Theory*. Chicago: University of Chicago Press.

May, R. M.
1973 *Stability and Complexity in Model Ecosystems.* Monographs in Population Biology, No. 6. Princeton, New Jersey: Princeton University Press.

Mayr, E.
1976 "The emergence of evolutionary novelties." In E. Mayr (ed.), *Evolution and the Diversity of Life*, Cambridge, Mass.: Harvard-Belknap Press.

Mercado, B. L.
1978 "Biology, Problems and Control of *Rottboellia exaltata L.f.*: A Monograph." *Biotrop Bulletin* No. 14. Bogor, Indonesia: BIOTROP, SEAMEO Regional Center for Tropical Biology.

Merrill, E. D.
1926 *A Discussion and Bibliography of Philippine Flowering Plants.* Manila: Bureau of Printing.

1945 *Plant Life of the Pacific World.* New York: Macmillan Company.

Monkhouse, F. J., and J. Small
1978 *A Dictionary of the Natural Environment.* New York: John Wiley and Sons.

Moran, E. F.
1979 *Human Adaptability: An Introduction to Ecological Anthropology.* North Scituate, Mass.: Duxbury Press.

Mueller-Dombois, D., and H. Ellenberg
1974 *Aims and Methods of Vegetation Ecology.* New York: McGraw-Hill.

Netting, R. M. C.
1977 *Cultural Ecology*. Menlo Park, Cal.: Cummings Publishing Company.

Nie, N. H., et al.
1975 *Statistical Package for the Social Sciences*. New York: McGraw-Hill.

Nieuwolt, S.
1966 "A Comparison of Rainfall in the Exceptionally Dry Year 1963 and Average Conditions in Malaya." *Erdkunde*, Band XX, Heft 3: 171-81.

Norbeck, E.
1956 "David P. Barrows' Notes on Philippine Ethnology." *Journal of East Asiatic Studies* 5(3):229-54.

Odum, E. P.
1971 *Fundamentals of Ecology*. Philadelphia: W. B. Saunders Company.

Opler, P. A., et al.
1977 "Recovery of Tropical Lowland Forest Ecosystems." In *Recovery and Restoration of Damaged Ecosystems*, J. Cairns et al. (eds.). Charlottesville: University Press of Virginia.

PAGASA (Weather Bureau, Philippines)
1966 *Annual Report of the Weather Bureau, FY 1965-1966*. Quezon City: PAGASA.

1966-72 *Philippine Annual Climatological Review for 1966-1972*. Quezon City: PAGASA.

1975 *Extremes of Temperature, Rainfall, Winds and Sea Level Pressure of the Philippines as of 1972*. Quezon City: PAGASA.

Paijmans, K. (ed.)
1976 *New Guinea Vegetation*. Amsterdam: Elsevier Scientific Publishing Company.

Parker, L.
1964 "Report on Work among the Negritos of Pampanga." *Asian Studies* 2(1):105-30.

Peterson, J. T.
1978 *The Ecology of Social Boundaries: Agta Foragers in the Philippines*. Illinois Studies in Anthropology, No. 11. Urbana: University of Illinois Press.

Philippine Commonwealth Weather Bureau
1903-38 *Annual Report of the Weather Bureau, 1903-1938*. Manila: Bureau of Printing.

Philippine Weather Bureau
1957-58 1960-65 *Annual Climatological Review for 1957-58, 1960-65*. Quezon City: Philippine Weather Bureau.

Pianka, E. R.
1978 *Evolutionary Ecology*. New York: Harper and Row.

Purseglove, J. W.
1972 *Tropical Crops: Monocotyledons*. London: Longman.

Rai, N. K.
1982 *From Forest to Field: A Study of Philippine Negrito Foragers in Transition*. Ph.D. dissertation, University of Hawaii, Department of Anthropology.

Reed, W. A.
1904 *Negritos of Zambales*. Ethnological Survey Publications. Manila: Bureau of Printing.

Reyes, M. R.
1959 "Natural Regeneration of the Philippine Dipterocarp Forest." *Philippine Journal of Forestry* 15(1-4):39-59.

Richards, P. W.
1952 *The Tropical Rain Forest*. Cambridge: Cambridge University Press.

Richerson, P. J.
1977 "Ecology and Human Ecology: A Comparison of Theories in the Biological and Social Sciences." *American Ethnologist* 4(1):1-26.

Ricklefs, R. E.
1973 *Ecology*. London: Nelson.

Robbins, R. G.
1968 "The Biogeography of Tropical Rainforest in S.E. Asia." In *Proceedings, Symposium on Recent Advances in Tropical Ecology*, Part 2, R. Misra and B. Gopol (eds.). Varanasi, India: International Society for Tropical Ecology.

Salisbury, R. F.
1975 "Non-Equilibrium Models in New Guinea Ecology." *Anthropologica* 17(2):127-48.

Sandoval, M. P., and F. B. Mamaril, Jr.
1970 *Hydrogeology of Central Luzon*. Bureau of Mines, Report No. 70. Manila: Bureau of Mines.

San Jose, J. J., and E. Medina
1975 "Effect of Fire on Organic Matter Production and Water Balance in a Tropical Savanna." In *Tropical Ecological Systems: Trends in Terrestrial and Aquatic Research*, F. B. Galley and E. Medina (eds.), in series titled, *Ecological Studies: Analysis and Synthesis*, J. Jacobs et al. (eds.). New York: Springer-Verlag.

Schebesta, P.
1952-57 *Die Negrito Asiens*, 3 Vols. Wien-Modling: St. Gabriel Verlag.

Schlegel, S.
1979 *Tiruray Subsistence*. Quezon City: Ateneo de Manila University Press.

Science Education Center, University of the Philippines
1971 *Plants of the Philippines*. Quezon City, Philippines: University of the Philippines Press.

Scotter, D. R.
1970 "Soil Temperatures Under Grass Fires." *Australian Journal of Soil Research* 8:273-79.

Serevo, T. S.
1949 "Some Observations on the Effects of Different Methods of Logging on Residual Stand and on Natural Reproduction." *Philippine Journal of Forestry* 6(4):363-81.

Shimizu, H.
1981 "Marriage and Bridewealth among Southwestern Mt. Pinatubo Negritos, Zambales--In Relation to the Composition of the Extended Family." *Ninzokugaku-kenkyu* [The Japanese journal of ethnology] 46(1):35-54.

1983 "Communicating with Spirits: A Study of the *Manganito* Seance among the Southwestern Pinatubo Negritos." *East Asian Cultural Studies* 22(1-4):129-67.

Shimizu, H., and M. Yamashita
1982 "Grammatical Analysis of Kakilingan Sambal, the Language of the Southwestern Pinatubo Negritos." *The Proceedings of the Department of Humanities, College of General Education, University if Tokyo* 76:115-65.

Simmonds, N. W.
1962 *The Evolution of the Bananas*. London: Longmans.

Slatyer, R. O. (ed.)
1976 *Dynamic Changes in Terrestrial Ecosystems: Patterns of Change, Techniques for Study and Applications to Management*. MAB Technical Notes No. 4. Paris: UNESCO.

Smith, W. D.
1909 "Contributions to the Physiography of the Philippine Islands: IV. The Country Between Subig and Mount Pinatubo." *Philippine Journal of Science* 4(A):19-25.

Spencer, J. E.
1966 *Shifting Cultivation in Southeastern Asia*. Berkeley: University of California Press.

Spurr, S. H., and B. V. Barnes
1964 *Forest Ecology*. New York: John Wiley and Sons.

Street, J. M.
1969 "An Evaluation of the Concept of Carrying Capacity." *Professional Geographer* 21(2):104-7.

Sutherland, J. A.
1971 *Introduction to Tropical Agriculture*. London: Angus and Robertson.

Tamhane, R. V., et al.
1959 "Effect of Intensity of Rainfall on the Soil Loss and Runoff." *Journal of the Indian Society of Soil Science* 7(4):231-38.

Tansley, A. G.
1920 "The Classification of Vegetation and the Concepts of Development." *Journal of Ecology* 8:118-49.

1939 *The British Islands and their Vegetation*. London: Cambridge University Press.

Taylor, C. N.
1936 *Odyssey of the Islands*. New York: Charles Scribner's Sons.

Temple, P. H.
1972 "Measurements of Runoff and Soil Erosion at an Erosion Plot Scale with Particular Reference to Tanzania." *Geografiska Annaler* 54A(3-4):203-20.

Temple, P. H., and D. H. Murray-Rust
1972 "Sheetwash Measurements on Erosion Plots at Mfumbwe, Eastern Uluguru Mountains, Tanzania." *Geografiska Annaler* 54A(3-4):195-202.

UNESCO
1978 *Tropical Forest Ecosystems*. Vendome: Presses Universitairs de France.

Vayda, A. P., and R. A. Rappaport
1968 "Ecology, Cultural and Noncultural." In *Introduction to Cultural Anthropology*, J. Clifton (ed.). Boston: Houghton Mifflin.

Vogl, R. J.
1977 "Fire: A Destructive Menace or a Natural Process." In *Recovery and Restoration of Damaged Ecosystems*, J. Cairns et al. (eds.). Charlottesville: University Press of Virginia.

Walker, D., and J. C. Guppy
1976 "Generic Plant Assemblages in the Highland Forests of Papua, New Guinea." *Australian Journal of Ecology* 1:203-12.

Wallace, B. J.
1970 *Hill and Valley Farmers*. Cambridge, Mass.: Schenkman Publishing Company.

Walter, H.
1971 *Ecology of Tropical and Subtropical Vegetation*. Edinburgh: Oliver and Boyd.

1973 *Vegetation of the Earth*. London: The English Universities Press, Ltd.

War Office (United States of America)
1908 *Report of the Philippine Commission*. Washington, D.C.: Government Printing Office.

Wernstedt, F. L., and J. E. Spencer
1967 *The Philippine Island World*. Berkeley: University of California Press.

Wharton, C. H.
1968 "Man, Fire and Wild Cattle in Southeast Asia." In *Proceedings, Seventh Annual Tall Timbers Fire Ecology Conference*. Tallahassee, Fla.

Whitford, H. N.
1906 "The Vegetation of the Lamao Forest Reserve." *Philippine Journal of Science* 1(4 and 6):373-429, 637-81.

1911 *The Forests of the Philippines*. Bureau of Forestry, Bulletin No. 10. Manila: Bureau of Printing.

Whyte, R. O.
1974 *Tropical Grazing Lands: Communities and Constituent Species*. The Hague: Dr. W. Junk b.v., Publishers.

Winterhalder, B.
1980 "Environmental Analysis in Human Evolution and Adaptation Research." *Human Ecology* 8(2):135-70.

Yamada, T., et al.
1975 "Effects of TFP on *Eulalia, Miscanthus sinensis* Anderss." In *Proceedings, Fifth Asian-Pacific Weed Science Society*. Tokyo, Japan.

Yen, E. E., and J. Nance
1976 *Further Studies on the Tasaday*. Panamin Foundation Research Series No. 2. Makati, Rizal: Panamin Foundation (Philippines).

Zinke, P. J., et al.
1978 "Soil Fertility Aspects of the Lua Forest Fallow System of Shifting Cultivation." In *Farmers in the Forest*, P. Kunstadter et al. (eds.). Honolulu: University Press of Hawaii.

Zuniga, J. M. de
1973 *Status of the Philippines in 1800*. Manila: Filipiniana Book Guild.